Rafael Pablo Massocato

Development of a robot for exploring hazardous environments

Rafael Pablo Massocato

Development of a robot for exploring hazardous environments

Hardware, microcontroller board and supervisory system

ScienciaScripts

Imprint

Any brand names and product names mentioned in this book are subject to trademark, brand or patent protection and are trademarks or registered trademarks of their respective holders. The use of brand names, product names, common names, trade names, product descriptions etc. even without a particular marking in this work is in no way to be construed to mean that such names may be regarded as unrestricted in respect of trademark and brand protection legislation and could thus be used by anyone.

Cover image: www.ingimage.com

This book is a translation from the original published under ISBN 978-613-9-70443-9.

Publisher:
Sciencia Scripts
is a trademark of
Dodo Books Indian Ocean Ltd. and OmniScriptum S.R.L publishing group

120 High Road, East Finchley, London, N2 9ED, United Kingdom
Str. Armeneasca 28/1, office 1, Chisinau MD-2012, Republic of Moldova, Europe
Printed at: see last page
ISBN: 978-620-8-21132-5

I dedicate this work to Alexandre and Magda, who are undoubtedly the best parents! And to Daniele for all her help throughout my undergraduate years.

ACKNOWLEDGEMENTS

I would like to thank all the people who first believed in me and my willpower, be they friends, parents, teachers or family. I would especially like to thank Professor Karl Phillip Pumhagen Ehlers Peixoto Dittrich Buhr for all the knowledge I acquired in programming classes, and for the concern he always had for our learning and development not only as academics, but as programmers. I would also like to thank the teachers involved in my final year work: Kleyton Hoffmann and Daniel Calixto Fagonde Moraes, who have always been following my work and helping with extremely important tips and guidance.

SUMMARY

The purpose of the collection robot is to help in situations where there is a threat to life, whether due to the suspicion of an explosive artefact, environments where there are flammable or toxic gases, allowing it to reconnoitre the environment using cameras and in some cases collect the material and take it to a safe place. As it is remote-controlled, it has a supervisory system, allowing it to handle suspicious material without putting human life at risk. The components to be used in the robot's manufacture were chosen and its main characteristics and limitations identified, as well as the necessary adaptations. Then all the robot's *hardware* was assembled and its constructive capabilities were assessed, such as locomotion, moving the camera and the robotic arm, as well as reading sensors. The robot's capabilities include identifying the presence of smoke and some flammable gases in the environment, whether the object collected is emitting heat and, using the camera, recognising the environment and interacting with it by controlling the robotic arm and moving around. The connection between the electronic components and the microcon- troller used in the robot was covered in detail, as was the construction of the chassis and the connection of all the components involved. All the *software* needed to control the robot's functionalities was also developed, as well as the embedded system responsible for driving the *hardware* and controlling it. Tests were carried out in a commercial room to identify the robot's capabilities, simulating a real danger situation, where the robot was able to transmit the reading from the gas and temperature sensors up to a maximum distance of 50 metres. During the simulations, images were obtained from the camera at a maximum distance of 25 metres using *Wi-Fi*. The robot travelled at a maximum speed of 1.33 km/h and the autonomy time was 23 minutes. In 66 per cent of the attempts, it managed to pick up and transport the object using the robotic arm.

Keywords: Collector robot, Microcontroller, Supervisory control.

SUMMARY

CHAPTER 1

INTRODUCTION

Robotics is becoming increasingly popular and accessible to everyone, whether for leisure or work, robots have been helping man for decades in the most diverse activities, whether it's assembling industrial nails, complex welding, ocean or space exploration. In the future, robots will be increasingly present in our lives. Since the space race, Nasa has sent robots to explore neighbouring planets, and around eight robots have been successfully sent to the planet Mars. Their aim is to explore the planets, identify their properties and whether life once existed on them. The same goes for the oceans, which are explored by surface-controlled submarines, through which new species of animals can be identified and the environment better understood. There are other tasks that put human life at risk, such as disarming bombs, radioactive environments, among others, in which robots prove to be very efficient. A robot has been built that is capable of accessing dangerous locations where there may be threats of explosions or flammable and toxic gases, allowing it to recognise the location and sometimes collect the explosive or flammable material and take it to a safe place. The robot was built to fulfil these needs without costing too much. Ready-made devices were chosen, which have already been tested and are easily found on the market; these devices must be adapted so that they can perform the functions required for the robot to work. An explanation is given to identify what a robot collector is, how they are classified and which type will be developed in this work. This is followed by a detailed explanation of the main components used, and finally the construction details, assembly and testing of all the robot's *hardware*. The registers used in the libraries developed for the embedded *software* responsible for controlling the robot are also presented, as well as the development of the supervisory system. The final stage consists of testing the collection robot in an environment that simulates a real danger situation, identifying its capabilities and limitations.

1.1   DESCRIPTION OF THE PROBLEM

Some environments can be fatal for human beings because they contain chemical, radioactive and other threats. Likewise, every time a specialist goes to a site where there is explosive material or flammable gases, their life is in danger. According to Reno (2000), the number of incidents involving explosive artefacts in the United States in four years exceeds two thousand. For this reason, collection robots are now widely used, helping specialists to take explosive artefacts to a safe place without endangering human life. In Brazil, 30 robots were purchased to help with World Cup security in 2014, totalling an investment of 16.2 million reais (CAPUTO, 2014). These robots are very expensive and there are no low-cost models available on the market, let alone national companies that manufacture them.

1.2   BACKGROUND

With the development of the collection robot, it is possible to access potentially dangerous places without putting human life at risk. It can check for the presence of flammable and toxic gases or collect explosive materials to be taken to a safe place. As the robots marketed for this purpose are imported, the purpose of this robot is to be a prototype for the development of a national product, taking into account the costs, as the equipment is often lost to damage from explosions or flames.

1.3   OBJECTIVES

Developing a remote-controlled robot to access dangerous sites where there is a threat of explosion, flammable or toxic gases, enabling it to carry out site reconnaissance or collect small explosive or flammable materials.

1.3.1   Specific objectives

- Develop a microcontroller board.

- Develop the embedded *software* to be used for control.

- Developing the supervisory system and communicating with the robot.

- Make general adjustments to the robot

- Carry out tests with the robot in a simulated hazardous environment to identify its capabilities and limitations.

- Finalise the project and present conclusions.

CHAPTER 2

THEORETICAL BACKGROUND

2.1 ROBOTS

According to Inovation (2005), located in the State of California in the United States, according to the documentation provided by the site: robots are machines that obtain information about the environment through sensors and use this information to follow instructions on how to act. However, it is becoming increasingly difficult to define what a robot actually is, as there are now many types of robot used in a wide variety of activities. This can be understood as the functionalities provided by a *smartphone*, such as geolocation or the use of mobile data, but it is known that not all robots have these functionalities despite being very advanced in other respects. According to Wise (2005), there is no exact definition of what a robot is, but they do have essential characteristics to determine whether or not they are a robot. In this way it is possible to identify when building a machine whether it really is a robot. Among the essentials are listed: sensors, movement, energy and intelligence according to what it has been programmed to do. The author's definition is therefore a system that contains sensors, control systems, manipulators, a power supply and *software* working together to perform a certain task. He goes on to say that designing, building, testing and programming a robot involves physics, mechanical engineering, electrical engineering, maths and computer science.

According to Inovation (2005), there are various types of robots with a wide variety of purposes. Over time, robots have changed industry and are bringing more and more comfort to our lives. They have increased productivity in factories and made many tasks safer and easier to carry out. For this reason, there are now many types of robots, some of which have been developed to carry out specific activities such as welding and assembly, while others, such as flying *drones*, can carry out the most varied tasks, depending on the accessories attached to them.

2.1.1 Explorer Robot

The first exploratory robot was developed by NASA in 1997, called So- joumer. According to the Nasa website (2015), its aim was to reach the surface of the planet Mars and transmit data in images. This robot was carried by the Pathfinder module and transmitted several photos of the planet, as well as the composition of rocks found in the Martian soil. With dimensions of 68 cm long, 48 cm wide and 30 cm high, the robot weighed around 10 kg. Its mission ended the same year when communication with Earth was lost. This mission was the second cheapest in NASA's history. Figure 1 shows the Sojoumer prototype.

Figura 1   - Sojoumer Explorer Robot

Source: Nasa (2015)

Many robots have been developed with the aim of exploring environments, including other explorers who have been sent on missions to the planet Mars. Currently there is still a robot called Opportunity, which according to Nasa (2015), in eleven years of exploration has covered more than 42 kilometres of Martian soil. Some robots have been developed for use in rubble removal, bomb defusal and other activities that are dangerous for humans. Some companies sell or rent this type of robot, but they are expensive.

1.1.2   Exploration Robot Components

According to the information provided by Nasa (2015), explorer robots have some essential components. Every robot is equipped with a body that can house all the components and withstand the mechanical stresses exerted on its parts; this would be the skeleton or chassis where everything is grouped together. According to Nasa (2015), it has a brain capable of processing the data collected during exploration, as well as the energy that powers its circuits and enables the robot to operate satisfactorily. An explorer-type robot is equipped with at least one camera in order to identify what is in front of it and to the sides and whether it can move in that direction, in addition to providing the

the ability to observe various important objects or any kind of threat (NASA, 2015). Comparable to human vision, it is through the camera that the robot will perceive all the visual elements around it. In the same way that the robot visualises with the camera, it can hear sounds through microphones, thus allowing it to identify noises that are of interest to whoever is operating it. Most of the time, according to Nasa (2015), an explorer robot also has a robotic arm that allows it to collect samples or activate a device as if it were a human arm, but obviously without the same precision of movement, as it lacks sensitivity. However, in many cases this imprecision can be overlooked, because even with sudden movements it will be able to press a button or

8

remove an obstacle from its path. In order to move around, a robot needs a device that can support its weight and still move forward. According to Nasa (2015), many robots use legs, others have tracks and the most common are those with wheels. Depending on the application, an explorer robot can have various accessories that allow it to expand its capabilities in solving tasks. These devices can be sensors of all kinds, torches, actuators or any other device that can be attached to it.

1.1.3   Collector Robot

These robots were developed with the aim of accessing dangerous locations where there are artefacts that could cause explosions. Their aim is to find the artefact and transport it to a safe place where it can be detonated or take it to containment chambers suitable for storing bombs (SKILLANO; KHATIB, 2015). There are various types of collection robots, used for military applications, public safety and by specialised bomb squads. These robots are a type of explorer robot, possessing several characteristics of explorers. Both are remote-controlled robots and their components are practically the same (FOUNDATION, 2015). According to Benson (2008), the first collecting robot was developed by Lieutenant Colonel Peter Miller in 1972. Fed up with the deaths of colleagues in bomb disposal, Peter developed his first model to carry out this task. It was a motorised wheelbarrow that was fitted with a hook capable of pulling vehicles containing explosive devices without putting the bomb disposal expert's life at risk. In recent years, collection robots have been greatly improved. According to Ryder (2005), they are capable of handling explosives built by specialists, as the movements performed by the robot have become increasingly smooth, thus allowing the explosive artefact to be handled more safely, without it exploding or, depending on its complexity, even being disarmed. Some companies, such as iRobotCorp, manufacture various types of exploration robots, including robots designed to collect explosive materials, which were widely used by US troops to disarm roadside bombs in the Iraq and Afghanistan wars, as well as helping in the Fukushima nuclear accident. Figure 2 shows the PackBot 510 robot produced by iRobotCorp and used by the US army.

Figura 2   - PackBot 510

Source: iRobotCorp (2017).

2.1.4   Cost of a Collector Robot

Collector robots are mostly manufactured by US companies and even the simplest models cost more than a sports car. The prices range from $30,000 to $72,000 for the models we researched, taking into account the capacity of each robot and their constructive properties. Table 1 lists the main commercialised robots and the approximate value of each one.

2.1.5   Collector Robot Components

A robot that has to carry explosive artefacts is a little more robust than an explorer robot, as the artefact to be moved can often be very large.

Table 1 - Collector robot budget

Manufacturer	Model	Value (US$)
Dr Robot Inc	Jaguar V4 with arm	30.000,00
SuperDroid Robors	Super-Droid H2-S	32.500,00
Icor Technology	Mini-Calibre	50.000,00
Robotex	Avatar III	50.000,00
iRobot	510 Packbot	72.000,00

Source: the author.

mass. This robustness is important to withstand small impacts while moving the artefact to a safe place. As with any robot, a brain is required, responsible for control and for transmitting or receiving data. This component must be as effective as that of an explorer robot, due to its importance in carrying out the tasks of a collecting robot. The vast majority of collecting robots have more than one camera in their system, allowing them to visualise the object to be collected from several different angles without even having to touch the object. The vast majority of today's explorer robots also use more than one camera (MANSOOR et al., 2001).

Every robot that is going to collect an explosive artefact needs a robotic arm to move towards the explosive charge and gently hold it so that it doesn't fall and accidentally explode along the way. Many exploration robots have extremely precise robotic arms capable of collecting and analysing rock samples. In the same way as exploration robots, the collection robot can move around on wheels. However, most models use tracks that are capable of overcoming much larger obstacles and climbing stairs, which are more efficient when they are double-jointed and articulated (MANSOOR et al., 2001).

2.1.5.1 Electric motors and gearboxes

Today, electric motors are widely used in industry in the most diverse applications; they first appeared in the 9th century, according to Sandin (2003), and are known for their ability to convert mechanical energy into electrical energy or vice versa. There are two types of motors, alternating current and direct current. According to Inovation (2005), the latter are those that require a power supply and use a non-variable voltage, i.e. they have two poles, one negative and the other positive. Direct current motors are powered by a current in which the poles do not vary; only to invert the rotation of direct current or DC motors is it necessary to invert the polarity.

This way it will rotate clockwise or anti-clockwise. Reduction gearboxes, as the name suggests, are responsible for reducing the final speed of the motor. According to Sandin (2003), this means that it turns its shaft more slowly at the output of the gearbox. Automatically, by reducing the speed, an inversely proportional physical quantity called torque is obtained, which is responsible for the force generated at the end of the shaft, for example making the motor capable of pushing a much greater weight than it could without the torque coupling. Figure 3 shows a motor with a gearbox. Motors are generally sold

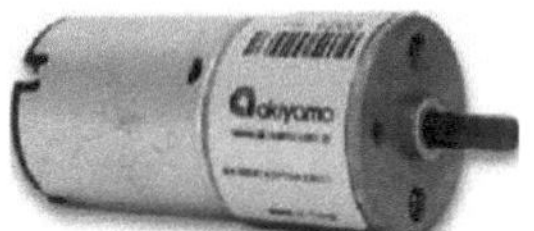

Figura 3 - Engine and gearbox

Source: Sandin (2003).

together with the gearbox, because in most applications not much speed is used and without the reduction provided by the gearbox, the force produced in the wheels would be insufficient to move a robot with its entire structure.

2.1.5.2   Transistorised board for PWM control

According to Sandin (2003), there are companies that manufacture the most diverse boards for assemblies involving devices driven by pulse width modulation, or PWM. These boards are used in aeromodelling, among other radio-controlled devices, and their main function is to control the speed of DC motors, which have high current consumption. The torque provided by the motor is proportional to the energy needed to lift it out of inertia.

The board chosen for this work is the DBH-12V, which can work with voltages ranging from 5 to 12 Volts and a maximum current of 30 Amps per channel. Figure 4 shows the board used. In addition to these advantages, the module weighs around 50 grams and can work with frequencies of up to 16KHz.

2.1.5.3   Servomotors

According to Sandin (2003), servo motors have the characteristic of turning precisely according to the signal they receive, unlike other motors where this is not possible.

Figura 4   - PWM control board

Source: XL Semi (2016).

control. There are various types of servomotors, but due to cost and speed the motors used in this project are DC motors. Figure 5 shows an example of a servomotor. The ability to rotate with a certain precision is provided by a

Figura 5   - Servo motor example

Source: Adapted from Inovation (2005)

circuit mounted inside the DC servomotor, which consists of a board with an integrated circuit and a potentiometer, the motor's rotation is controlled according to the signal received and the resistance presented in the potentiometer, according to Inovation (2005).

2.1.5.4 Chassis and suspension

The chassis is the structure that supports all the equipment and devices housed in the robot. All the other parts are attached to it, which is why it is so important. According to Siegwart and Nourbakhsh (2004), they are normally made from light, resistant material that does not deform much. Their thickness determines the amount of weight they can support. They can be made from materials such as acrylic or other polymers for small robots and aluminium or steel for medium and large robots. According to Adams (1992), the suspension is attached directly to the chassis and is responsible for

for absorbing the impacts suffered by the wheels, making sure that as many of these impacts as possible are absorbed without affecting the chassis. There are various types of suspensions available for use and the automotive industry is increasingly researching and developing new concepts capable of generating more comfort and reducing the impacts suffered by the structure even more. There are two types of suspension used: independent and semi-rigid axle. In axle suspensions, the wheels tilt as the obstacle hits the wheels, causing one to rise and the other to follow part of the movement in the opposite direction, shown in Figure 6 in (A). This type of suspension is not very suitable for use on wheels that exert traction, because if one of the wheels is suspended the other will reduce the vehicle's stability considerably. Independent suspensions, on the other hand, as the name implies

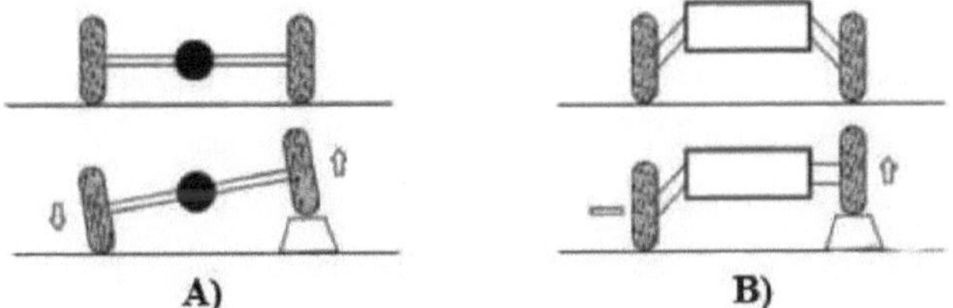

Figura 6   - Example of axle suspension (A) and independent suspension (B)

Source: Adapted from Adams (1992)

suggests, each wheel is capable of suffering and absorbing impacts individually, as shown in Figure 6 in (B). According to Adams (1992), this allows the robot, when crossing an obstacle, to stay firmly on the ground when the wheels hit it.

1.1.5.5   Batteries

There are currently many different types of batteries, for the most diverse applications and with the most different voltages and charging capacities (BRAUNL, 2003). Because of this huge variety, only the battery used in this project will be discussed. Known as Li-Ion, it is widely used by all of us, as most notebooks, mobile phones and digital cameras on the market use this type of battery in association. Its main characteristics include its high charge capacity and light weight. Although they have various working voltages, the ones discussed here, known as the 18650 model, work with a voltage of 3.7 V and can supply an average of 2 A/h of current. According to Braunl (2003), by combining the batteries in series, it is possible to increase the voltage of the batteries so that they can be used in the same way.

used by motors and circuits, in the same way that associating the batteries in parallel will allow the current supplied to be greater, thus increasing the available load. It is possible to use both forms of association, obtaining both a higher voltage and a higher current.

2.1.5.6 Voltage regulator boards

Voltage regulator boards are used to raise or lower the voltage supplied to a circuit, thus allowing batteries associated with voltages other than those used in the circuits to power them. Most of these boards are powered by direct current supplied by batteries and have circuits that allow the output voltage to be regulated according to the project's needs. The board used in this project has a variable input voltage of 4 to 38 Volts and a variable output of 1.25 to 36 Volts, capable of supplying a current of up to 5 Amps. According to the manufacturer XL Semi (2012), the integrated circuit used in the board is the XL4005. Its main features are its small size and simplicity of use, with only five pins, which are used for input, output, grounding, a switch and a feedback pin that allows the output voltage to be regulated. Mounted on a board, the XL4005 circuit has a multiturn potentiometer that allows the voltage to be adjusted at the output of the board. Figure 7 shows the board used.

Figura 7   - Regulating plate XL4005

Source: XL Semi (2012).

2.1.5.7 Raspberry Pi computer

According to Pi (2005), Raspberry Pi is a very small computer, developed by the foundation of the same name in the UK in early 2012. It can be used in a variety of applications due to its small size and low power consumption. It doesn't have a monitor or inbuilt input/output devices; instead, it has several inputs and outputs that allow you to connect devices such as a keyboard, mouse, monitor and so on.

It is capable of surfing the Internet and playing videos in high definition, and usually runs Linux operating systems. According to Pi (2005), one of the main features of the Raspberry Pi is that version 3 brings innovations such as Wi-Fi 802.11 and Bluetooth 4.1 and uses a BCM2837 processor with 4 ARM Cortex-A53 cores running at 1.2GHz, as well as 1GB of volatile memory. The Raspberry's purpose is to be a computer like any other, but it has GPIO pins that can be used to connect sensors and actuators, as well as to communicate with other devices. Figure 8 shows the Raspberry Pi 3

Figura 8   - Raspberry Pi 3 Model B

Source: Raspberry Pi Foundation.

Model B+ has a camera input next to the HDMI output, four USB ports, a micro-USB power port, a low-resolution display output, an HDMI output for high-resolution video, audio output, ethemet connection and a memory card slot (not shown in the picture) on the underside of the card.

2.1.5.8   Raspberry Pi camera module

The camera developed by Adafruit for use with the Raspberry Pi has two different versions, which can be connected directly to the Raspberry via the CSI (serial camera interface) input used for this purpose. Version 2 allows a resolution of up to 1080p at a rate of 30 frames per second and a density of 8 Megapixels. The main feature of this camera is its size (around 2.5 cm wide by 2.5 cm high) and light weight (around 3 grams). As it is a module, it is mounted on a board and connected to the Raspeberry via a *flat* cable. There is also a parallel camera that can be used in conjunction with the Raspberry and costs less. Figure 9 shows the camera mounted on the board.

Figura 9   - Camera module

Source: Raspberry Pi Funda- tion.

2.1.5.9   Atmel ATmega2560 microcontroller

According to Wise (2005), there are currently many microcontrollers from a wide variety of manufacturers, and it is possible to find a microcontroller on the market with the characteristics needed to develop a prototype or any other project. According to Pereira (2007), a microcontroller is a device with several input and output ports. These ports can be used to connect sensors to read a signal or switch on an LED, for example. This happens according to the configuration made in its *firmware*. According to Pereira *(2007)*, *firmware* is *software* that is run inside an embedded system in order to carry out a certain programmed task. For example, executing the steps for washing clothes in a machine for this purpose. According to Atmel (2017), there are several microcontrollers sold by Atmel in its line known as megaAVR, and in this project the ATmega2560 model is discussed. It uses an 8-bit AVR core, running at a speed of 16 MHz. It also has a 256 Kb flash memory (for storing code to be executed), 8 Kb RAM memory (to be used for storing temporary data) and 4 Kb ROM (for storing boot data). It has one hundred pins, including: digital inputs/outputs, PWM (Pulse Modulation), UART, I2C and SPI communication, among other functions distributed across its 11 *ports. NQ-SQ* Figure 10 shows the Atmel microcontroller that will be used in this project. Note that this microcontroller is SMD encapsulated and must be mounted on the surface of the board.

2.1.5.10   SMD to DIP adapter board

The small size of SMD components makes it impossible to use them in *protoboards* and other circuit prototyping tests. Therefore, in order to access

Figure 10 - Atmega2560

Source: Atmel (2017).

Due to its small size, an adapter is used to distribute the pins of the integrated circuit in a socket of the size used in DIP models. Figure 11 shows the board used; it is interesting to note that it supports SMD integrated circuits from 32 to 100 pins (BRAUNL, 2003).

Figure 11 - SMD/DIP adapter
Source: Hayoui Eletronics (2017).

2.1.5.11   Xbee Explorer USB Adapter

To make communication between the microcontroller and the computer possible, it is necessary to use a board whose main function is to interface the XBee modules with the USARTs of the computer and the microcontroller. The main chip used in this board is manufactured by FTDI. According to the manufacturer, the FT232R is available in a 28-pin package and the entire USB protocol is managed by the device. The characteristics of the circuit make it possible to work with voltages of 3.3 V used by the Xbee modules and 5 V present in the USB ports. The board used in this project has four *LEDs* connected to the pins that receive and transmit the data (RX and TX), making it possible to identify when data is travelling between the USB and the XBee. Figure 12 shows the board used in this project.

Figura 12   - FT232R module

Source: FTDI Chip (2017).

2.1.5.12   Sensors

According to Murphy (2000), their main function is to capture information from the environment so that it can be transformed and interpreted by a wide variety of electronic devices. It is through them that we know what the temperature or humidity is in a given location, whether a driver has been drinking alcohol, or whether it is necessary to switch on the lights of a lamppost because it has become dark on an avenue. Many sensors will not be useful in this project, so only the temperature and smoke sensors are covered. According to Mazidi (2014), the MQ-135 smoke sensor is one that identifies certain toxic gases in the environment in which it is located, being able to identify the presence of various gases such as ammonia, carbon dioxide, benzene, nitric oxide, as well as smoke or alcohol. It is important to emphasise that this is a simple sensor that does not have the ability to identify the concentration of each of the gases, but rather their presence in a certain concentration range. The manufacturer's manual specifies a reading accuracy of between 10 ppm and 1000 ppm. This sensor is shown in Figure 13 in A, mounted on a module. According to Miyadaira (2013), the purpose of the DS18B20 temperature sensor is to determine whether it is cold or hot, obtaining this value from the air, or depending on the type of sensor, it can also identify the temperature of the material it is in contact with, be it solid or liquid. Generally, for liquids, the sensor looks like a metal rod attached to a wire and for solids, it can be very simple and look like a transistor, such as the LM35, manufactured by National Semiconductor. The DS18B20 sensor used in this project, according to the manual provided by the manufacturer Integrated (2017), can measure temperatures ranging from - 55 °C to 125 °C, with an accuracy of plus or minus 0.5 °C and resolution

Figure 13 - Gas and temperature sensors
Source: Maxim Integrated (2015).

up to 12 bits. This sensor is shown in Figure 13 in B.

2.1.5.13   Robotic arm

Robotic arms are currently widely used in industry for welding and assembling parts. The first ones appeared with the Industrial Revolution and have been increasingly perfected (MURPHY, 2000). With the increase in precision, there are robotic arms that, according to experts, are capable of building bridges entirely by themselves, just by welding to the structure (INNOVATION, 2005). The robotic arm used in this project is much more modest, since these state-of-the-art devices are expensive; and the aim of this work is for it to be able to perform simple tasks, such as removing a small obstacle or pressing a button. This device has four axes of rotation, allowing it to move 180° with a twist, i.e. clockwise and anti-clockwise, as well as three joints as if it were a shoulder and elbow and the hand, compared to the human arm. At the end of the arm is a claw capable of grasping and holding small objects. Servomotors are used to carry out the movements of the arm joints, which are activated by a pulse of modulated length, supplied by a digital output (SANDIN, 2003). Depending on the time at high level received, the servomotor will act in a certain number of degrees, most often 0° to 180°. The gripper will work in the same way, connected via gears to the servomotor it will open or close according to the pulse received. Figure 14 shows the model of the arm in question, which already has the servomotors attached as well as the claw. The wires should be arranged in a harness that does not restrict the mobility of the arm in all its ranges of movement.

Figure 14 - 6-axis robotic arm
Source: Sandin (2003).

2.1.5.14 Supervisory

Supervisory *software* is *software* that runs on a computer and is generally used to monitor and control *hardware* devices. According to Inovation (2005), it is possible, for example, to identify whether a generator is working properly, how much energy is being generated, the temperature of the components involved, as well as countless other quantities that can be measured and monitored.

Using control *software*, it is also possible to control a robot using a computer or mobile phone, so that it performs certain actions according to the controller's wishes. Through this, data can be sent or received, depending on the chosen means of communication, be it wired or wireless. Figure 15 shows an example of an Android supervisory system for controlling the PackBot robot developed by IRobot.

2.1.5.15 Data communication via Wi-Fi

Wireless data communication, which began in the 9th century with the communication of a wireless telegraph in Europe, has evolved to the present day where almost all devices have not just one wireless communication, but several of them. According to Wise (2005), the most widely used today are Wi-Fi, GPRS and Bluetooth, as well as, of course, satellite data transmission, whether for georeferencing or other types of data transmission and reception. The type of communication used between the robot and the supervisory system using Wi-fi technology will be discussed here, where the notebook will have to connect to the robot via a computer.

Figure 15 - Example of a supervisory system
Source: IRobot.

of a wireless network. According to Molloy (2011), *Access Points* are wireless connection points between other wireless or wired devices and, depending on the case, a Gateway, if necessary to access the Internet. There are various AP models, some of which allow signal replication to increase power in places where the *Access Point* has a weak signal. There are many different manufacturers of signal repeaters. In Brazil, this equipment needs to be approved by Anatei before it can be sold on the market.

2.1.5.16   Data communication via Xbee

XBee is the name of a family of compatible modules used for wireless data transmission, manufactured by the company Digi International, with the most diverse powers and applications. One of the main features of these modules is their practical connection, using only Vcc with a voltage of 3.3V, ground and two other pins used for communication: for sending (Tx) and receiving (Rx) data.

The modules can come with a number of different aerials depending on the needs of the project and the desired signal strength. Another characteristic of these modules is their low data transmission rate of no more than 250 Kbps in most models. There are several models of XBee modules that are compatible with each other. The models used in this work are the XBee Series 2, shown in Figure 16.

Figura 16   - XBee module

Source: Digi International Inc.(2009).

According to Digi (2014), the great advantage of using these modules is that their range is very significant in terms of the energy consumed by the devices. They are able to send data at a low speed, but their energy consumption is also low given the range of the signal, and they operate at an ISM frequency of 2.4 Ghz.

2.1.5.17 Xbox 360 controller

The Xbox 360 controller is used in the video game produced by Microsoft. According to the manufacturer, it can be connected to a computer's USB port. Figure 17 shows an Xbox 360 controller, which has eleven buttons (4 to 10, plus X,Y,A,B) and three directional controls (two of which are analogue) that send around 9 different signals each. These are generally used for movement purposes, as they allow you to work on at least two axes each.

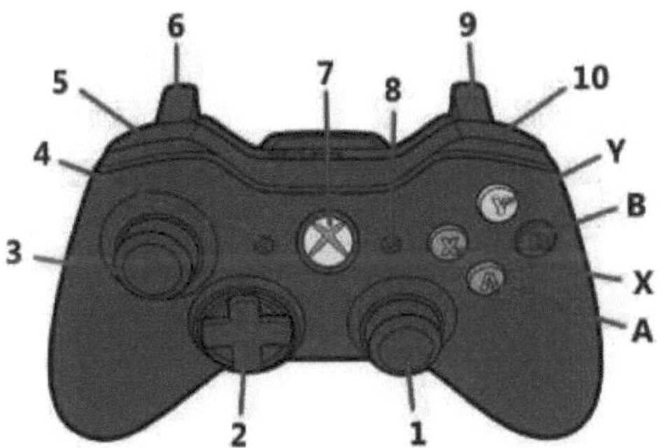

Figure 17 - Xbox 360 controller

Source: Microsoft Xbox.

According to Microsoft (2017), some of the buttons have what is called a context-sensitive interface, so the signal sent by the button can be higher or lower.

This can help with tasks that require greater precision and sensitivity. The accessory is fully compatible with the Windows platform and when a button is pressed an event is generated that can be handled by the operating system, thus allowing interaction with games and other devices that can be controlled by the computer.

2.1.5.18 Programming languages

The programming languages used to develop the work are C and C#. According to Deitei and Deitei (2011), the C language was created in 1972 by Dennis Ritchie. It is a structured language that was created for the development of operating systems, later becoming very popular and being compiled for the most diverse platforms, including embedded systems. The C# language is a compiled language developed by Microsoft and is based on C++ concepts, such as object orientation, one of the main inherited characteristics (DEITEL; DEITEL, 2006). As it is a compiled language, it requires a compiler, which is often included with the development IDE, which is the set of tools that allows the programmer to organise, compile and *debug* the project if necessary. To carry out this project, we chose to use the Integrated Development Environment called Visual Studio, which is the main one for use with .NET tools provided by Microsoft (DEITEL; DEITEL, 2006). According to Waslawick (2011), the modelling language used for the development of this work is UML (Unified Modelling Language), which can be used to describe things, and through which diagrams will be created to understand how the *software* should be developed, at the same time as the documentation is drawn up.

CHAPTER 3

METHODOLOGY

The research carried out is of the quantitative-qualitative type and the research modality is of the experimental type. To carry out this work, the method adopted is hypothetical-deductive, where hypotheses are raised in a general way and then particular data is analysed (MARTINS, 2000).

The complete system, to be developed in this work, is shown in Figure 18. Communication is carried out between the supervisory system and the collecting robot, allowing control and manipulation of actions. The supervisory system displays the data collected by the robot, such as images and the temperature sensor reading. This data makes it possible to identify whether the collected material is increasing in temperature, thus helping to identify changes in the material while it is being collected (MANSOOR et al., 2001).

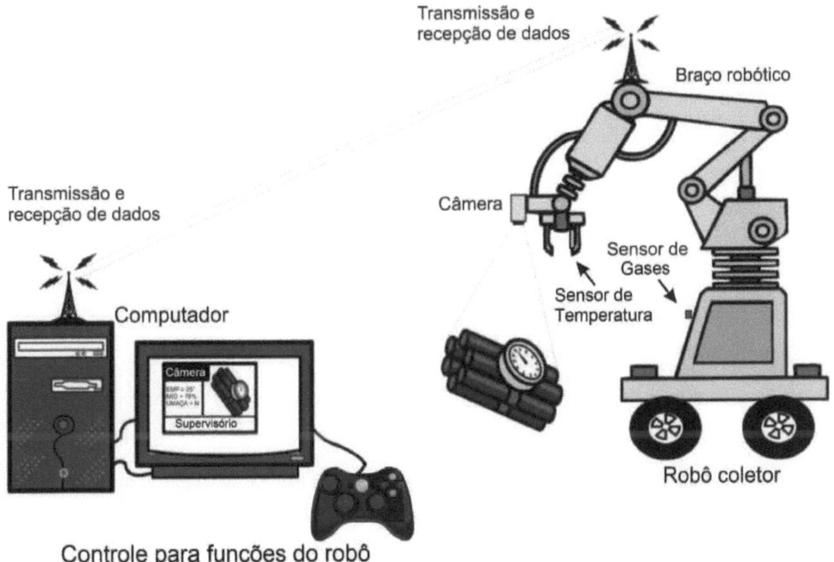

Figura 18   - Complete system with robot collector and supervisor

Source: Adapted from Wirelles Bomb Disposal Robot

18.1 DESIGN AND CONSTRUCTION OF THE ROBOT CHASSIS

18.1.1 Robot CAD design

A model developed with the CAD tool made it possible to virtually draw the components used in the robot, helping to design and visualise its structure. Using this model, the materials used in the robot's structure were chosen and the components were positioned so that their capacities would be most efficient. A special place was designed on the front of the robot to attach the robotic arm; this space proved to be suitable for the movements of the arm's joints.

Initially, the chassis was designed, as this is where all the components are attached. The design of the chassis, which should have had a trapezoidal shape when cut from the side, was changed to a more rectangular model, making the suspension and motors occupy a large useful area inside the chassis. The main reasons for the change were the high cost of the motors and gearboxes and improving the robot's centre of gravity, allowing for greater stability when moving on slopes. Throughout the construction of the chassis, changes were made to fix the suspension, the possibility of installing larger wheels and fixing the camera. Figure 19 shows the initial design of the robot's chassis.

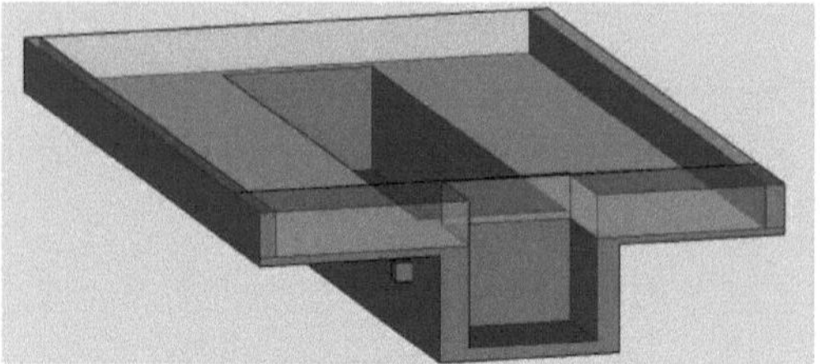

Figura 19   - Robot chassis

Source: The Author.

The first part of the project to undergo changes was the robot's suspension. Due to the size and positioning of the shock absorbers and springs, it was necessary to build an extension to the chassis, allowing the suspension to be attached at three points, two at the front housing the motors and one at the rear. It is through the suspension that the robot is able to overcome certain obstacles more easily. Aluminium arms and a coil spring system were used, along with an oil shock absorber, as designed.

The motors and gearboxes were attached to the upper part of the suspension, unlike the design where the motors were at the bottom; a cylindrical part was manufactured to house the motors and allow them to be moved. Figure 20 shows the initial design of the suspension system, motors and wheels.

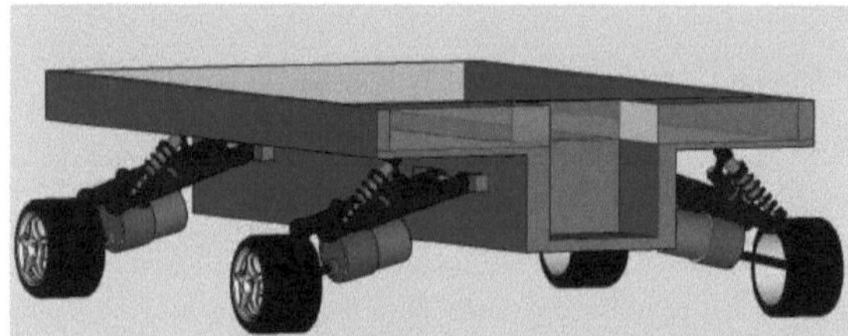

Figura 20   - Suspension system, engines and wheels

The next step was to attach some electronic components so that it would be possible to assemble the circuits needed for the robot to operate and the batteries to power the circuits. Figure 21 shows the inside of the robot, where all the electronic components must be attached.

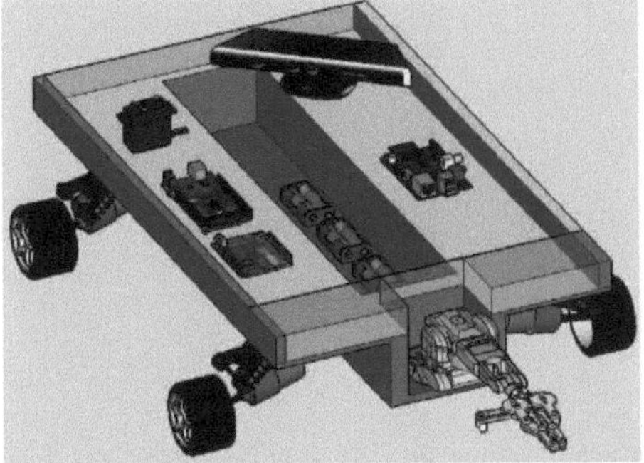

Figura 21   - Electronic components

In the last stage, an acrylic cover should close off the top of the robot, making it look nicer and protecting the circuits from moisture and dust. The version proposed for assembly is shown in Figure 22.

In practice, the development of the chassis didn't just change its shape, but was adapted as needs arose. The suspension system needed five assemblies to work as shown below.

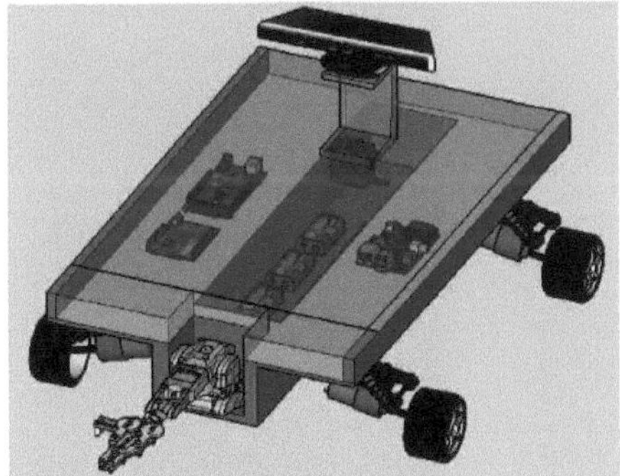

Figure 22 - Completed robot prototype

3.1.2 Chassis mould

The material chosen to build the chassis was fibreglass, due to its low cost and easy maintenance, as well as other interesting properties such as resistance and malleability. Initially, a mould made from wood, wood glue and fabric was developed, using which the shape of most of the chassis was developed. The mould is shown in Figure 23.

Figure 23 - Chassis mould

Source: The Author.

Once the mould was ready, a layer of fabric was added to allow the resin and fibre to sit on top of it, thus avoiding the whole area having to be closed off with wood, which would slow down the work and increase the weight of the chassis unnecessarily. Figure 24 shows the second stage.

Figure 24 - Chassis mould with fabric

Source: The Author.

3.1.3 Chassis manufacture

Once the mould was complete, two layers of resin and fibre were added to add strength to the mould. The next steps included adjusting the parts and applying plastic putty. After the entire chassis was sanded, the parts were painted and the suspension was ready to be assembled, the result is shown in Figure 25.

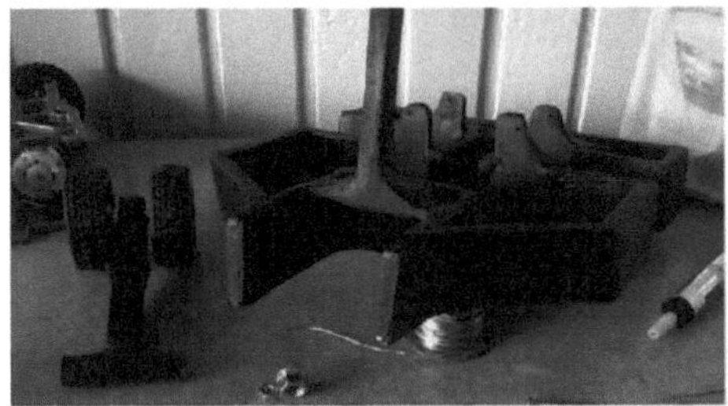

Figura 25   - Chassis for mounting the suspension

Source: The Author.

3.2 MICROCONTROLLER BOARD DEVELOPED WITH ATMEGA2560

For the development of the microcontroller board used in this project, three versions of the board were printed and corroded. The first two versions had very thin tracks and even though they were arranged equidistant from each other, they ended up coming together during printing on the board. The tracks were 0.4mm wide, which is the same width as the pins on the ATmega2560 microcontroller.

To make the board, we consulted documents made available by the manufacturer Atmel, as well as articles related to the use of this microcontroller and other board projects we found. The circuit developed with the Eagle tool is shown in Figure 26, after which all the details of the circuit used are explained. Among the main

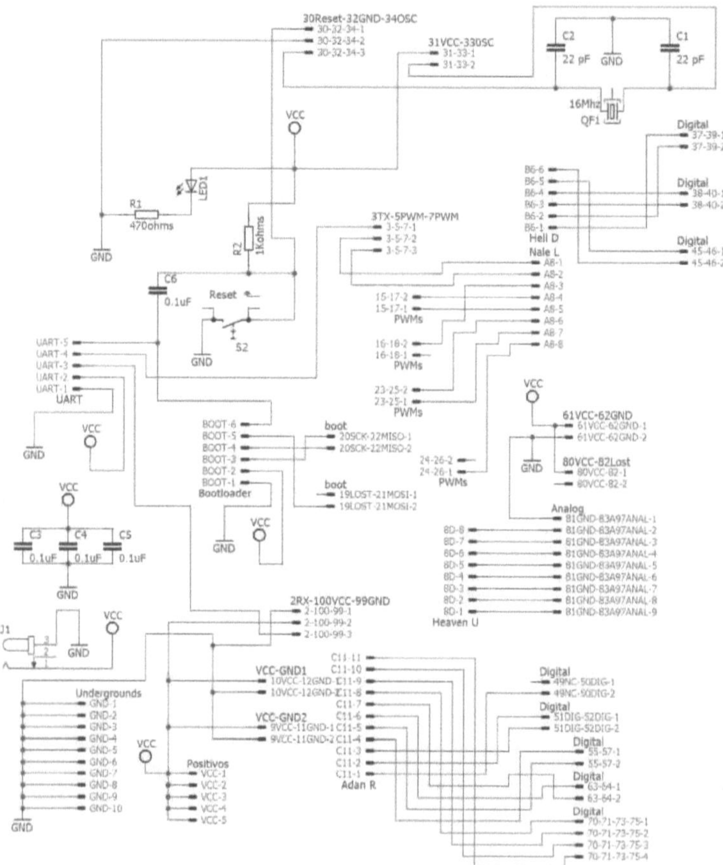

Figura 26   - Microcontroller board circuit diagram

Source: The Author.

pins related to the project we can mention those connected to Vcc: 10, 31, 61, 80 and 100, as well as the pins connected to Ground: 11, 32, 62, 81 and 99. Between these pins the *datasheet* provided by the manufacturer recommends the use of 3 capacitors as a *bypass* filter, with values of 0.1 uF, filtering the signal received by the microcontroller.

avoiding possible AC noise and other voltage variations across the board. Other widely used pins are those responsible for communication via UART: Ground, Vcc, RXO, TXO and a connection to the *Reset* which has a 0.1 uF capacitor connected in series, as instructed by the manufacturer.

3.2.1 Oscillator crystal circuit

According to Atmel (2017), to use a ceramic crystal *clocked* at 16 MHz, it must be connected to pins 33 and 34 of the microcontroller and a 22 pF capacitor must be connected in parallel with each of these pins and ground. This part of the circuit is shown in Figure 27.

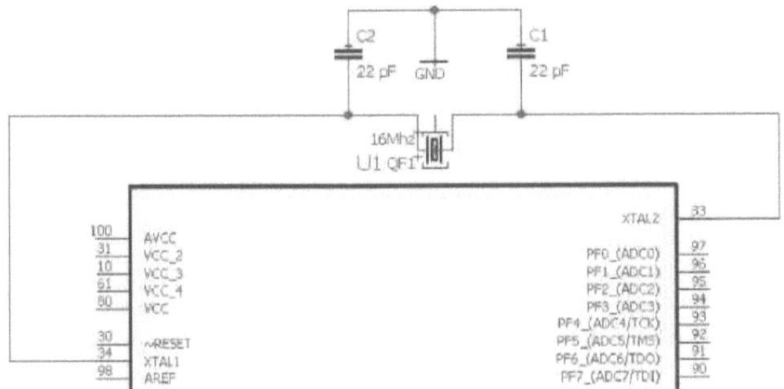

Figure 27 - Oscillator crystal circuit

Source: The Author.

1.1.2   Two-layer board design

The board was developed using the Eagle tool, which allows the circuit to be drawn with the appropriate components. After making all the component connections in the circuit, the board was routed and files were then generated and printed on the board representing the circuit developed with the software. Figure 28 shows the lower side of the board in A and the upper side in B.

1.1.3   Plate corrosion

To corrode the phenolite plate and leave only the tracks that are part of the circuit, a solution of *FeCLft*, diluted in water with a concentration of 30 to 40 per cent, was used. The plate was immersed in the solution and turned over so that the corrosion

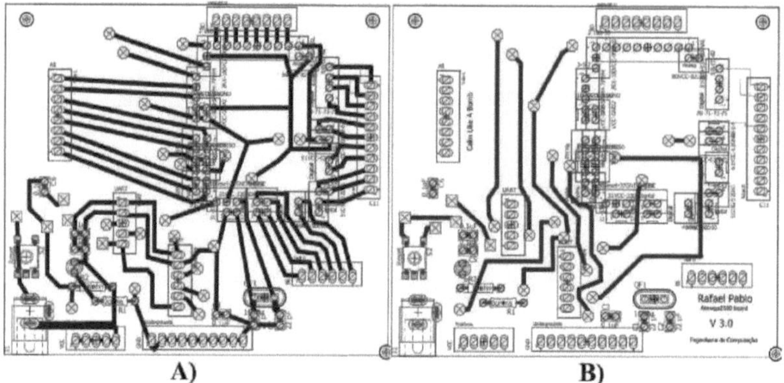

Figure 28 - Drawing of the top and bottom layer of the developed plate
Source: The Author.

This process is shown in Figure 29, and after about ten minutes the plate was completely corroded.

Figure 29 - Corrosion of the phenolite plate

Source: The Author.

1.1.4   Welding components

The components were soldered onto the board. Because it is a two-layer printed circuit, there are several points where the tracks start on one side of the board and end on the other. It was possible to position all the components so that they were on the same side of the board, such as capacitors, resistors and connectors. Many *jumpers* were used to allow the tracks to change sides along the board, demonstrating that more layers would be needed to achieve superior quality. The ATmega2560 integrated circuit was the component that involved the most time during soldering. As it is an SMD component and contains 100 pins with a small size, it had to be mounted on the adapter and then fitted to the board. Figure 30 shows the board after soldering the components on the lower left and upper right sides.

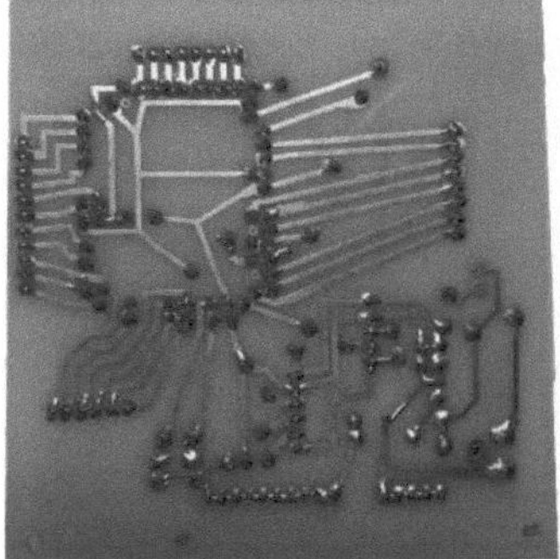

Figure 30 - Soldering the board components

Source: The Author.

3.2.5 Testing the microcontroller board

A number of tests were carried out in order to ascertain the functioning of the board, the first of which was the continuity test applied to all the tracks and points on the board, to make sure that current was passing through, mainly due to the 25 existing *jumpers*. The pins used in the circuit were also tested, from the operation of digital ports with PWM to the reading of analogue ports, as well as communication via USART using RX and TX, these being pins 2 and 3 respectively.

3.3 ROBOT CIRCUITS

Every robot has various electronic components that must be connected or configured in the right way for it to work satisfactorily. According to Inovation (2005), these components are connected and form a circuit that varies in complexity according to the robot's capabilities, so the more tools it has, the larger its electrical circuits will be.

Communication between the Raspberry Pi and the microcontroller board is via an XBee connection, but there are also serial connections between the Raspberry Pi and the connected camera. The vast majority of connections are on the microcontroller board's ports, which are distributed to the other devices available on the robot. Each servomotor has two wires for power, as do the sensors and other boards. The voltages can vary according to need, but the connections for these devices are necessary.

To make the work easier to understand, the circuits are divided into sections for each of the components; the complete circuit can be found in Appendix A.

3.3.1 Raspberry Pi connections

The Raspberry Pi, being the highest processing board, will have many connections, including: camera input and data transmission via Wi-Fi. This board will also be connected via the USB port to the microcontroller board developed. The main objective is to supply the board with a voltage of 5 volts. Figure 31 shows the connection diagram for the Raspberry Pi device.

Figure 31 - Components connected to the Raspberry Pi

Source: The Author.

1.1.2   Connections of the microcontroller board developed

The board developed using the ATmega2560 microcontroller will control the motors responsible for moving the robot, reading the sensors and moving the camera. Due to the number of devices involved, this circuit has been divided into sections to make it easier to understand.

3.3.2.1 Servo circuit   to move the camera

Two servomotors are connected to the PB4 and PB5 ports on the microcontroller board, which are responsible for rotating the Raspberry Pi's camera in two axes, allowing the robot to view the front from about 200 degrees horizontally and 150 degrees vertically. The PB4 and PB5 ports can be configured as PWM outputs to control the servomotors. Figure 32 shows the circuit used to move the camera.

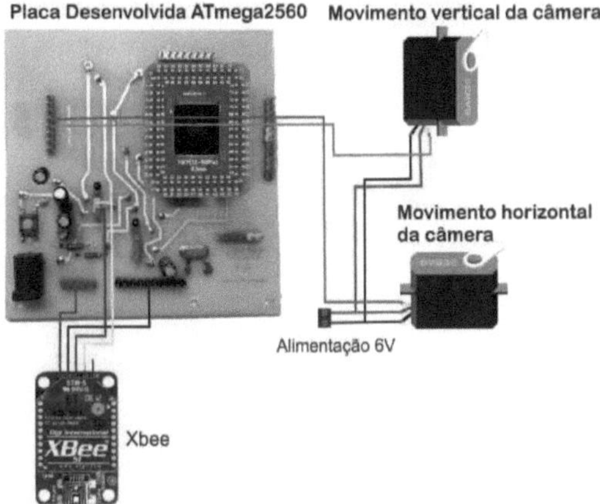

Figure 32 - Circuit used to move the camera

3.3.2.2   H-bridge circuit - Modification for PWM control

The motors responsible for moving the robot are connected to an H-bridge module, allowing the motors to be driven in two directions and, as it is transistorised, it is possible to vary the speed of each one. The digital outputs responsible for controlling the H-bridge are connected to pins PB6, PB7, PE4 and PE5, controlling the signal from each motor separately. This system is very common in wheeled robot circuits and basically works by sending a signal to drive the motor in a certain direction and a PWM signal to vary the speed. Figure 33 shows the circuit used in this project. It was also necessary to add a relay module, responsible for controlling the power supply to the H-bridge module while the system is not fully operational. This measure was adopted because sometimes the robot started to move as soon as its circuits were switched off.

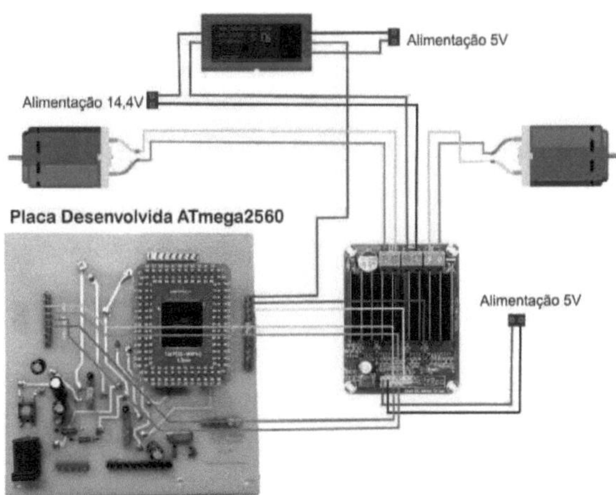

Figure 33 - H-bridge circuit

powered. Previously, the H-bridge in this project used a module with 4 relays, which were replaced by a transistorised module, thus allowing not only the direction of the motors to be controlled, but also their speed to be varied. This measure was adopted to make the robot's control more precise and efficient.

3.3.2.   3Robotic arm circuit

The components related to the robotic arm are also connected to the microcontroller board. These servomotors are responsible for actuating the joints or opening and closing the gripper, depending on the signal pulse they receive. The ports chosen for these connections were PG5, PH3, PH4, PH5 and PH6, which can be configured to emit a PWM signal. The circuit used for the robotic arm can be seen in Figure 34.

1.1.3   Device power supply

The power supply circuit requires several different voltages and powers, since each component used in the robot has very different tasks.

Eighteen 18650 batteries were used, each 3 of which were connected in series to provide a voltage of 11.1 V, which in turn were connected in parallel, increasing the current supply capacity to 13.6 Ah. The H-bridge was connected directly to the batteries, providing a voltage of 11.1 V for the motors.

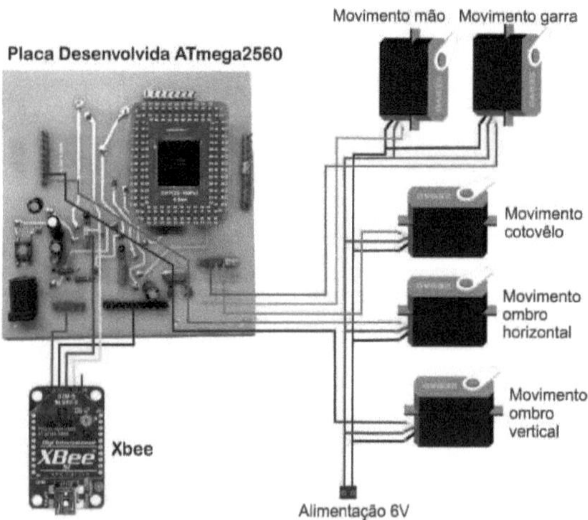

Figure 34 - Circuit used in the robotic arm
Source: The Author.

Two XL4005 regulator boards were used to regulate the voltages as required by the circuits, one set to

5 V to supply the Raspberry Pi and the microcontroller board developed, and the other to supply the servomotors with 6 V. The circuit used to power the devices can be seen in Figure 35. In addition to the

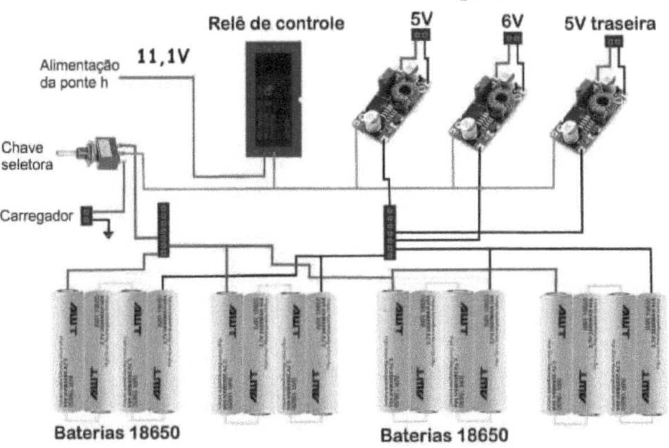

Fonte: O Autor.

Figure 35 - Power supply circuit for the robot's components

A two-position, three-terminal switch was used to control the power supply, powering all the robot's devices or leaving the circuit isolated for recharging the batteries. For recharging the batteries, an aluminium bracket was made, which houses the input connector for the power supplied by a 12 V supply for recharging the batteries.

3.3.3.   1Sensors

The temperature and gas sensors were installed in strategic locations on the robot to allow for better measurements. The DS18B20 temperature sensor was attached to the robot's gripper so that the reading is taken from the material that is in contact with it. The gas sensor was attached to the acrylic on the front of the robot, so that the air to be measured is always in the direction in which the robot is positioned. The sensors were both

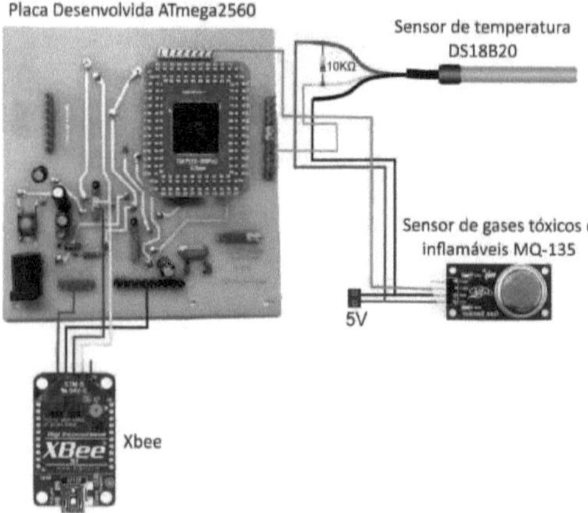

Figure 36 - Sensor circuit used by the robot

Source: The Author.

The temperature sensor uses a digital port for reading and the gas sensor (which allows both analogue and digital reading) was connected to a port for analogue reading using the AD converter.

3.4 ASSEMBLY

3.4.1 Suspension and engine assembly

After painting the parts, the chassis was drilled and it was realised that it would not be possible to attach the front suspension due to the size of the shock absorbers and springs.

To solve this problem, a more extensive fibre mould was made to fix the suspension, allowing it to function properly. The rear suspension also had flaws in the balance structure, which was reinforced with two more layers of fibre and resin.

The front suspension was assembled using two aluminium scales, fixed to the underside of the chassis, and two engine mountings that had been fabricated, where the engines fitted. The shock absorbers and springs were then mounted on two threaded rods installed in the upper part of the chassis.

3.4.2 Assembling   the robotic arm

The robotic arm was assembled with only four joints. Due to the arm's lateral movement, its weight increased considerably, which considerably reduced its performance. An extension was fitted to the gripper to make it easier to hold objects where the temperature sensor was installed.

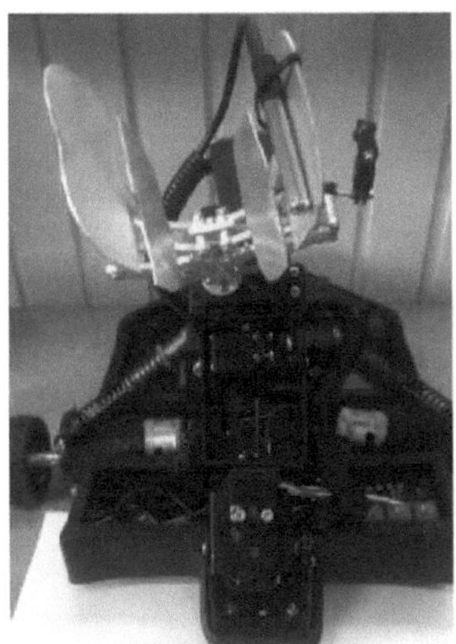

Figure 37 - Robotic arm

Figure 37 shows the assembly of the robotic arm, in which 5 servomotors and two screws were used to fix it. The motor cables were bundled in a harness to keep them together and better protected.

3.4.2.1 Adding another axis to the robotic arm

An additional axis was added to the robotic arm, allowing it to make lateral movements that were previously performed only by the robot's wheels, so that it could position itself towards the object to be collected.

3.4.3 Mounting the camera bracket

The camera for the Raspberry Pi was fixed to the top of the support and to house it we had to make a fibre box and fit a servomotor inside the cylinder. This servomotor is responsible for rotating the camera vertically and a bracket left over from the robotic arm was attached to it. Assembling the acrylic plate with the lower servomotor, the

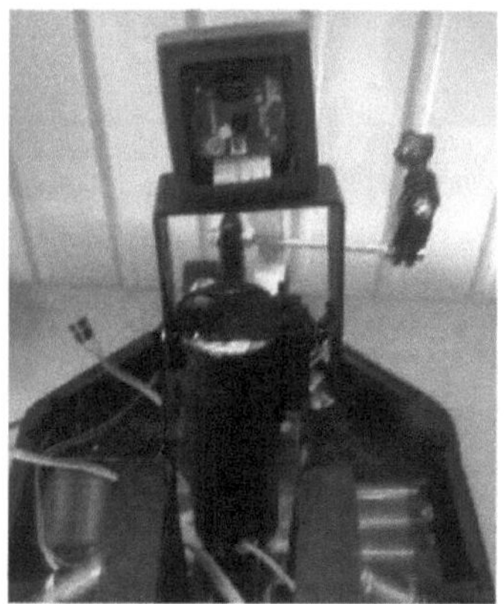

Figure 38 - Camera module

The cylinder with the upper servomotor and the camera housing mounted on the support is shown in Figure 38.

3.4.4 Assembling the acrylic parts

The acrylic pieces were cut and fixed onto the chassis, allowing an extra level for fixing the boards, sensors and also the servomotor, which is responsible for rotating the camera horizontally. The sides were fixed to the lower parts using angle brackets and screws, which ended up not fitting the chassis. The solution was to adjust the chassis to perfectly fit the acrylic parts, to which the Raspberry Pi was attached.

3.5 *HARDWARE* DOCUMENTATION

3.5.1 Context and requirements diagrams

The requirements needed to create the *software* have been collected and based on them it is possible to develop the *hardware* context diagram, which indicates which components are used and the technologies involved. Chart 2 shows this diagram, where the operator will control the robot via the supervisory system, sending and receiving data from the network connected to the Raspberry Pi, which will also be receiving commands and transmitting images. The Raspberry Pi will be exchanging data with the microcontrollers so that all the robot's functions can be carried out, such as moving and actuating the robotic arm.

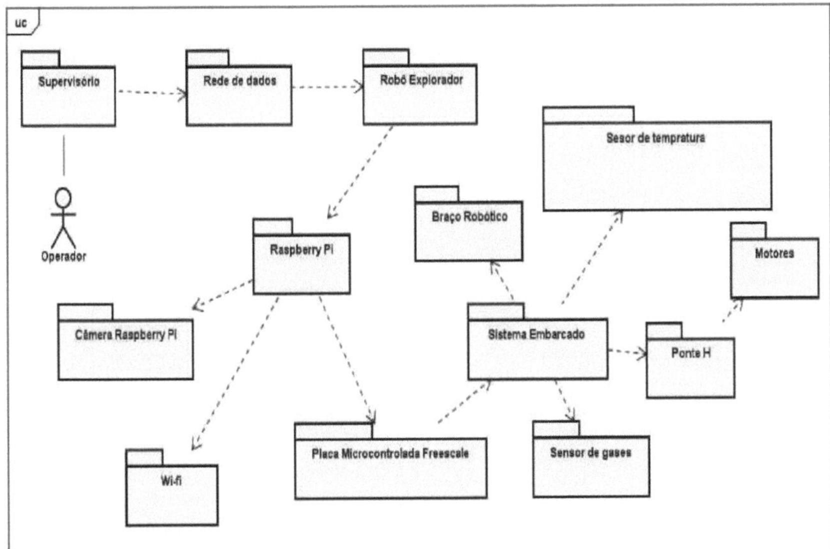

Table 2 - *Hardware* Context Diagram

Source: the author.

The *hardware* design diagram shows the functional requirements involved in the physical part of the project. As shown in Chart 3, the communication *software* will connect to the Raspbery Pi and the microcontroller to carry out the robot's required functions. In each of the frames you can see what is being managed and at the bottom what is desired as a result; each of the frames is connected according to the existing links in the robot's *hardware*.

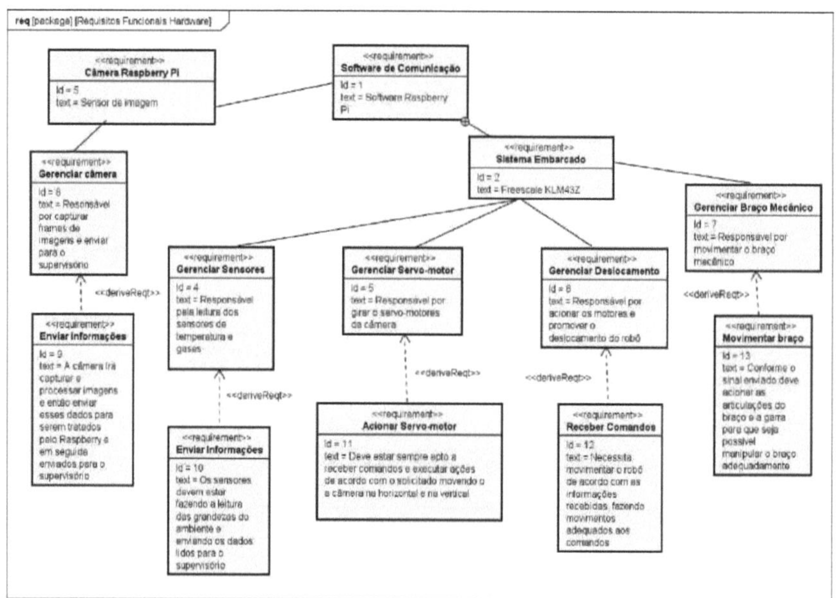

Table 3 - Functional *Hardware* Requirements
Source: the author.

3.6 *SOFTWARE* DOCUMENTATION

3.6.1 Identification of Operations and System Queries

The system's operations and queries have been mapped out and are shown in the diagrams involving the software's main functions, which are: connecting the microcontroller board, the camera and the Xbox controller. Chart 4 shows the sequence required to connect to the microcontroller board, where pressing the "Connect Serial" button calls the *Thread* responsible for managing the connection to the microcontroller board. Chart 5 shows the sequence required to establish the connection with the Raspberry Pi's camera, where pressing the "Connect Camera" button calls the *Thread* responsible for managing this connection. The diagram showing the sequence needed to establish the connection with the Xbox controller is shown in Box 6, where pressing the "Check Controller Connection" button displays in the field immediately to the right of the button whether the controller is "Connected" or "Disconnected". The system's queries and operations are those that call certain methods present in the *software*.

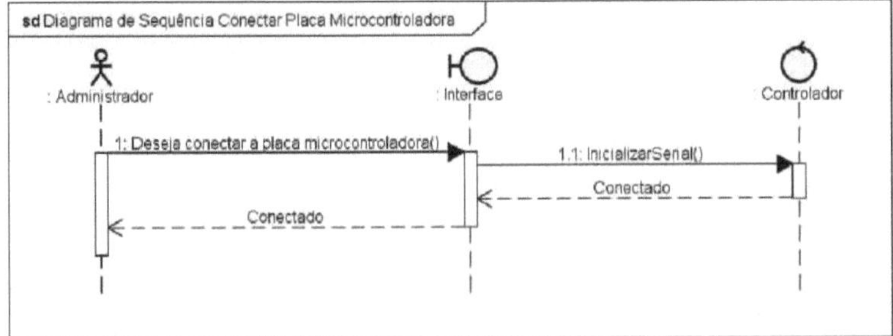

Chart 4 - Sequence diagram for connecting to the microcontroller board
Source: the author.

Box 5 - Sequence diagram for connecting to the camera
Source: the author.

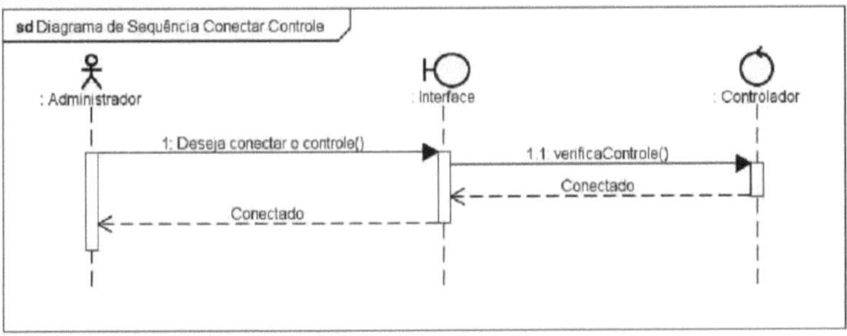

Chart 6 - Sequence diagram for connecting to the Xbox controller
Source: the author.

Table 1 - System operations and queries

System operation
updateSerial()
initialiseSerial()
closeSerial()
form(object: sender, EventArgs e)
xbox_cClick(object: sender, EventArgs e)
connect_serial_button_Click(object: sender, EventArgs e)
camera_video_button_Click(object: sender, EventArgs e)
stop_video_button_click(object: sender, EventArgs e)

Source: The Author.

These are tasks such as connecting to a device or activating an actuator. Queries are those that should result in a list of items previously saved in the software or a database; this supervisory does not use this concept. The main system operations are shown in Table 1.

3.6.2   Conceptual Model

The conceptual model provides a minimum outline of the attributes that the *software* should contain. The minimum expected are those shown in Table 7, which illustrates the Conceptual Model developed.

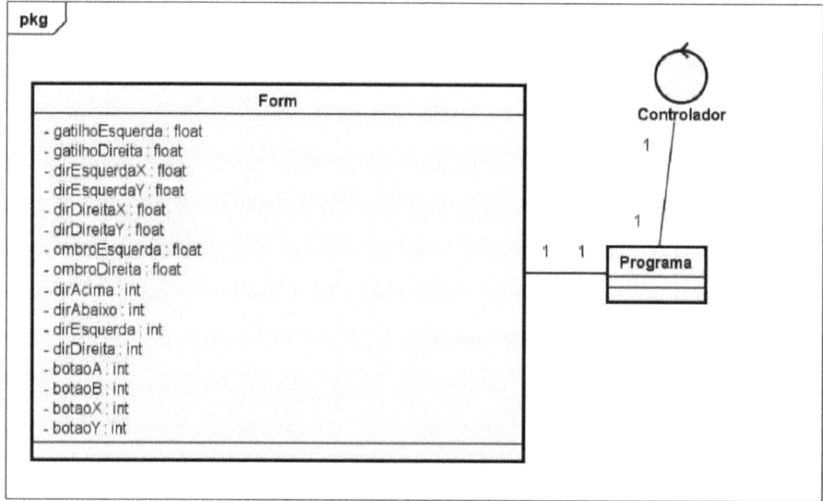

Chart 7 - Diagram of the Conceptual Model

Source: the author.

3.6.3   Project class diagram

The design class diagram, as its name suggests, shows which classes are developed in the software and how they relate to each other. It is similar to the design of a database where the tables are related to each other and each of them holds important data that may or may not be related to other tables. The design class diagram is shown in Box 8.

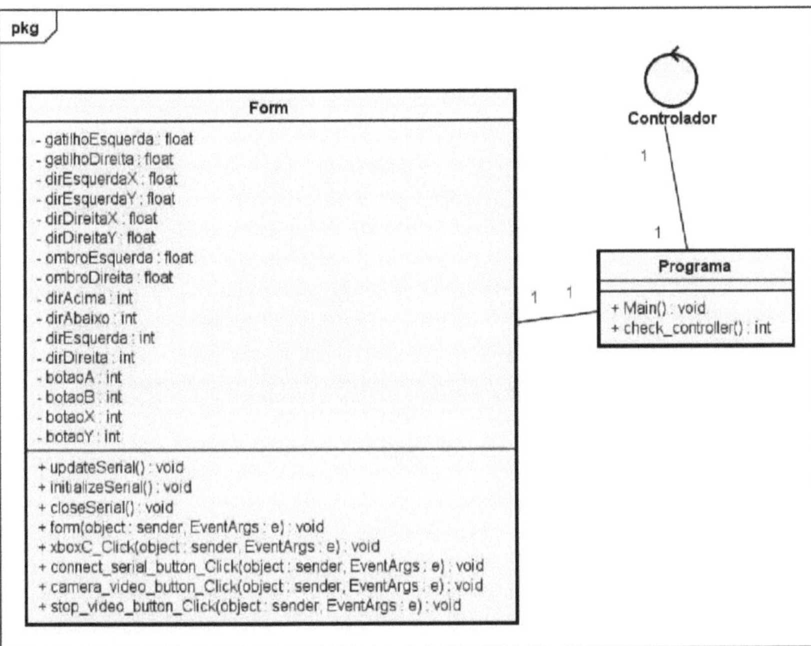

Chart 8 - Project class diagram

Source: the author.

3.6.   4Graphic design of   the centre screen

The *software*'s centre screen *is* made up of three main groups: The connection via serial port, connection to the camera and testing the Xbox controller buttons. The aim is to keep it simple, easy to use and to display only the essential functionalities for controlling the robot and obtaining information about its environment. The graphic design of the centre screen is shown in Figure 39.

Figure 39 - Supervisory centre screen
Source: the author.

3.6.5 Embedded *software* development

Based on the documentation available on the Atmel ATmega2560 microcontroller website and manual (2017), libraries were created using the registers responsible for each of the functionalities required to use the ports available on the board. These libraries are included in the solution being developed in order to use their functions, thus making the code cleaner and more organised. Libraries are usually created for reading a sensor, a digital output or the pulse width modulation used by servo motors.

3.6.5.1 Library for controlling IO ports

To use one of the microcontroller's ports, the pin names are first checked on the pinout diagram, always followed by PORT and the corresponding number. A function is then created to determine whether this pin will be an input or output, according to the parameters received by each of the registers involved. The registers involved in utilising the ports are: PORTx used to change the values of the outputs, DDRx responsible for indicating whether the port will be output or input, according to the value received 0 or 1 respectively, PINx input register used to read values on the pins. The x in each of the registers must be changed by the PORT to be configured and an 8-bit hexadecimal or binary number is passed as a parameter.

All the registers mentioned above must be configured so that the port chosen is a digital output or input to be used in the microcontroller. Two functions have been created in this library, one for input and one for output, with a signature in which it receives a variable of type *char* and another of type *int to* set the registers appropriately, indicating which PORT is to be configured and the corresponding pin.

44

3.6.3.1   Library for reading analogue ports

In order to use the analogue/digital converter present in the ATmega2560 microcontroller, three steps must be completed: the number of samples received must be determined according to the *clock* pulses, the read *buffer* is then disconnected by dividing the samples and finally the values are converted to a digital value. As the output of the analogue/digital converter is 10 bits, the registers involved are: REFS1:0 for selecting the voltage source used for reading, ADLAR for setting the most significant bit to the right or left, MUX3:0 for selecting the analogue channel used for reading, ADEN enables the analogue-to-digital converter, ADSC starts the conversion, ADPS2:0 determines the division of the processor clock to the ADC *clock*.

Three functions have been created in this library, one that does not receive parameters and initialises the converter configuration, another that indicates which PORT and pin are to be used as analogue inputs and one that reads the signal and controls the registers involved, in which case a value from the *upofloat* is returned in its signature.

3.6.3.2   Library for servo motor control

In order to control the servo motors, a library was developed capable of varying the value of the pins as a digital output, generating a pulse width modulated digital wave known as pwm. As the servos used work with a *clock* pulse of 20 ms, in addition to setting the registers we worked with this period to determine the changes in the wave that would allow an accuracy of 1°.

The registers involved in creating this library are: TCNTO which controls the *timer* count receives various parameters via 8 configuration bits, the TIRFO register determines the *flag* interrupts, plus the OCROA and OCROB registers which are used to compare the outputs.

Three functions were created, one responsible for configuring the ports as outputs, one to configure the desired values for the PWM and another to vary the values in degrees, according to the parameter passed in the method signature.

3.6.3.3   Library for USART communication

The last library developed is used to receive and send data to the microcontroller board, where sensor values can be read or command characters can be sent to execute an action already programmed in the *firmware*.

For USART configuration, the registers involved are: UBRRO which determines the speed of data transmission, UCRSOC which determines on which edge sampling is used, the size of the transmitted data and other bits relating to parity and stop bits are configured in the UCSROB and UCSROC registers respectively.

Three functions were created, one of them responsible for configuring all the parameters involved in sending and receiving data via USARTO, used by the microcontroller's serial port, and two others responsible for sending and receiving data, the one responsible for transmitting the data in its signature being two *float* values, referring to each of the sensors read. The function that receives data via the serial port receives a *string*, which is divided into several parameters.

3.6.6   Supervisory Software Development

The supervisory software was developed entirely using the C# language and some libraries written for it, such as XInputDotNetPure used to handle the events to be captured by the Xbox controller or IO used to make the connections involving the computer's serial port and the Wi-Fi network. The application contains only two classes, one of which is the main class and the other is responsible for monitoring the events sent by the Xbox controller. Two *Threads* were also created, which are responsible for updating the serial and the Xbox controller while the *software is in* use.

3.6.6.1   Connection to Raspberry Pi via IP

The data connection to the Raspberry Pi is via Wi-Fi and uses the HTTP protocol to connect to the camera located on the robot. Through an Apache server running on the Raspberry Pi, the camera images are transmitted over the network and can be captured by a device that connects to this server. The software loads the video address into a *stream* (a video transmission), with a *buffer* (memory used for loading) of a defined size, then the data that was in the *buffer* is loaded into an image that is displayed on the supervisory system. Appendix B shows an extract from the code used to initialise the connection to the camera on the Raspberry Pi.

3.6.Ó.2 Connection to the microcontroller board via USART

The serial connection to the microcontroller board uses functions from the IO library. Firstly, the serial port is checked to see if it is open, then values are read and written to the serial in the form of *strings* that are processed by the supervisor and *the* board's *firmware*. According to the *strings*, commands are sent or values are read, according to a specific format separated by commas.

The initialisation of the *thread* responsible for processing the serial is shown in Appendix C, which also shows the control of the serial connection and a data reception check.

3.6.6.3 Xbox controller event detection

To detect Xbox control events, methods were created to check when buttons are pressed on the control and when they are released. Modifiers were also developed for the analogue controllers and triggers to switch between linear, square and cubic modes, allowing movements to be made with less speed and greater precision when using the controller in the supervisory system. Basically, an object of the *GamePadState* type is instantiated and the methods check the positions of the analogue controls, trigger values and whether the other buttons have been pressed. There are many examples with the most varied libraries of how to implement these methods.

Due to the lack of precision displayed by the keyboard, the choice of a *joystick* proved necessary, as it is possible to vary not only the intensity with which the keys are pressed, but also their precision. Many remote-controlled robots have the option of *joystick* control. An excerpt from the code used to initialise the *thread* used to process the Xbox controller events, as well as the instantiation of an object of the *Gamepad* type and the detection of events for some buttons are shown in Appendix D.

3.6.7　Communication between control, supervisor and robot

For the system to function fully, there is integration between the control system, the supervisory system and the robot. The parameters chosen for the serial connection between the devices were 57600 bps, with 8 bits for sending data, 1 *stop* bit and no parity bit. These settings were used both on the supervisory side and on the microcontroller and likewise on the Xbee modules.

The control is connected via USB to the supervisor (in 1), and the data is sent serially via a specific port, allowing the computer to interpret each of the values sent by the control (in 2). The supervisor then reads each of these values using a serial connection component (in 3). The data is then passed on via another serial to the transmitting Xbee (in 4), which in turn replicates the signal to the receiving Xbee (in 5). The data is received by the other Xbee module (in 6), which sends the data via the microcontroller's serial port 1 (in 7) and then processed to carry out certain tasks. Communication in the opposite direction takes place in the same way for reading the sensors, with the exception of items (1 and 2, as the data does not return to the control). Serial communication is shown in Figure 40.

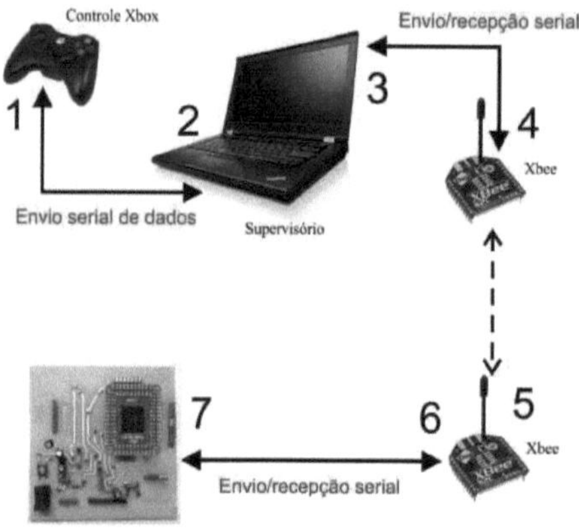

Figura 40   - Communication: Control, supervisor and microcontroller

Source: the author.

The data packets sent from the supervisor to the microcontroller are assembled in the following format: 'L,R,B,VAL,PAM,TIL,Y', where L and R are the values for driving the motors, B is the joint of the robotic arm that is moving, VAL is the value of the angle at which the chosen servo should move, PAM is the angle of the camera's horizontal servo and TIL is the angle of the servo responsible for the camera's vertical movement; the Y character is only used for control. The packets sent from the microcontroller to the supervisor have the following format: 'A,TEM,GAS,X', where the supervisor checks the serial looking for the character A, when it finds it, it reads the values until it finds X, which is the character used as a control to end the data reading. In this same step, the TEM and GAS fields are assigned the temperature and gas concentration, respectively, displayed on the supervisory screen.

The images transmitted by the Raspberry Pi use the following route to be read by the supervisor: the images obtained by the camera during reading are sent via serial to the Raspberry Pi via a cable/tar (in 1), these images are captured by a small piece of software (in 2), which is responsible for making them available via an HTTP address defined on an Apache server (in 3), where the supervisor makes a connection, via an *access point* (in 4), to capture this data and display it on a *picture box* component (in 4). The data transmission between the Raspberry Pi and the supervisory system is shown in Figure 41.

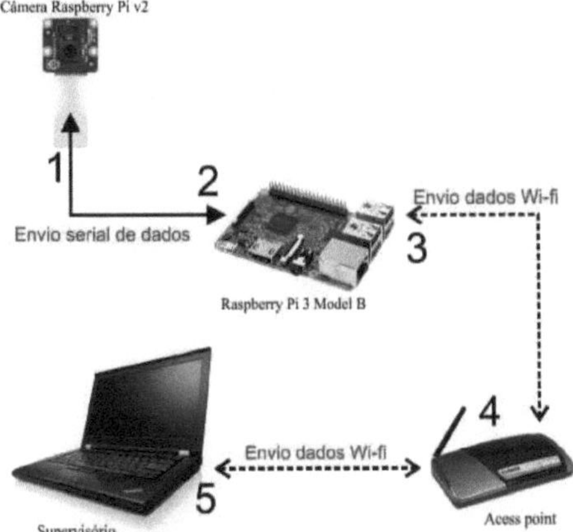

Figura 41   - Communication: Raspberry Pi and supervi-
solarium

Source: the author.

3.7 TESTING

After assembling the *hardware*, tests began to check the robot's capabilities and solve any problems that were encountered. The robot's main functionalities were tested, including locomotion, reading sensors, obtaining images from the camera, moving the robotic arm and battery life.

3.7.1 Locomotion

To move the robot, a library was developed to be used in the embedded *software* to move the robot. The main movements include moving forwards, backwards and turning left and right, with the variation presented by the sensitivity of the buttons on the Xbox 360 controller and the PWM control board improving the robot's locomotion precision exponentially.

A ramp was set up to check the maximum gradients that the robot could support when travelling. Due to the diameter of the wheels, the angle of entry into obstacles proved to be small, allowing climbs of up to 15°. Figure 42 shows the ramp set up for testing.

Figure 42 - Ramp for locomotion test

The lateral inclinations showed values close to 40°, without the robot's stability being compromised, while the exit angle presented the same problem as the entrance, causing the robot to lock up at values above 15°. Bearing in mind that the

If the robot is used for locomotion on flat ground, the values were acceptable. To determine the robot's maximum speed, a tape measure was extended and the time taken to cover 5 metres was approximately 13.55 seconds. Using this value, it was possible to determine the average scalar speed of 1.33 Km/h using Equation 3.1.

$$Vm = \frac{\Delta S}{\Delta t} \tag{3.1}$$

3.7.2 Sensor reading

Two *software* libraries have been developed to enable the board to read the sensors, one of which takes a digital reading (of the temperature sensor) and the other an analogue reading (of the gas sensor).

The reading of the DS18B20 temperature sensor was simulated by heating the robot's gripper with a squirrel. This sensor has a special feature that involves the use of a *pull-up* resistor to correctly identify the values.

A cigarette lighter was used to simulate the smoke sensor. The MQ-135 gas sensor is capable of reading smoke and toxic gases, including butane gas, which is used as a lighter fluid. According to Machado (2014), the value considered dangerous was obtained from the values considered dangerous

by bodies related to occupational health and safety, some gases in concentrations greater than 100 ppm already pose a risk to health and life. Figure 43 shows the values read from the supervisory screen

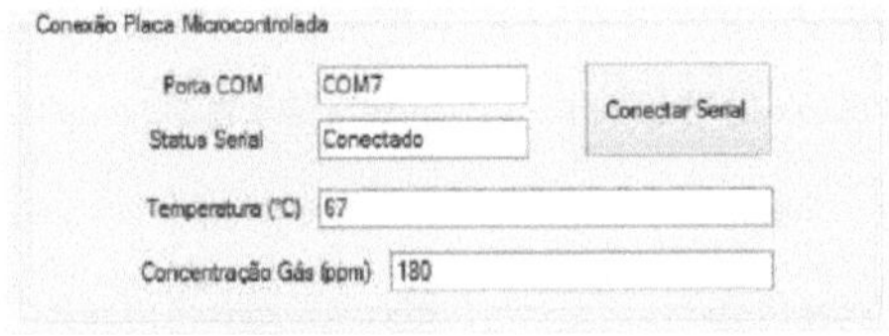

Figure 43 - Sensor reading in the Supervisory System

Source: The Author.

with the tests carried out to read the sensors. Once the tests had been carried out, the sensors were installed in the robot. The smoke sensor was attached under the acrylic on the front of the robot, leaving the sensor exposed and the board protected. The temperature sensor was attached to the robotic arm's gripper so that it could capture the temperature changes of the material being handled in the best possible way.

3.7.   3Robotic arm

The robotic arm performed as expected, but in order to add more torque the 996R servo used at the base of the arm was replaced by another with twice the torque capacity, the model chosen being the JX PDI-6221MG. The other joints proved strong enough to lift the arm, but in this model of robotic arm the arrangement of the servo motors, which are distributed along the arm, means that part of the force that could be used to lift the load is lost by lifting the weight of the motors themselves. Figure 44 shows the final assembly of the robotic arm, now with 4 axes.

Figure 44 - Robotic arm with 4 axes

Source: The Author.

The robotic arm proved capable of lifting an object weighing up to 150 grams and with maximum dimensions of 5 cm wide, 5 cm high and 25 cm long.

3.7.4   Signal Range

The power of the Wi-Fi signal emitted by the Raspberry Pi operating in router mode is similar to that of a router with an internal antenna, with power ranging from -78 dBm to - 82 dBm at a distance of 20 metres. The test was carried out with the signal passing through a masonry wall. At a distance of 20 metres, the signal strength was sufficient for transmitting and receiving data at a speed of 11 Mbps, in a range from -76 dBm to -86 dBm, which was considered acceptable (but a little slow). Figure 45 shows an example of a signal measurement reaching -68 dBm, using the Wifi Analyzer *software*, an Android application that measures Wi-Fi signal strength.

In order to improve the data connection, 2 XBee modules were used, which allowed the robot to be controlled from a much greater distance (reaching approximately 50 metres).

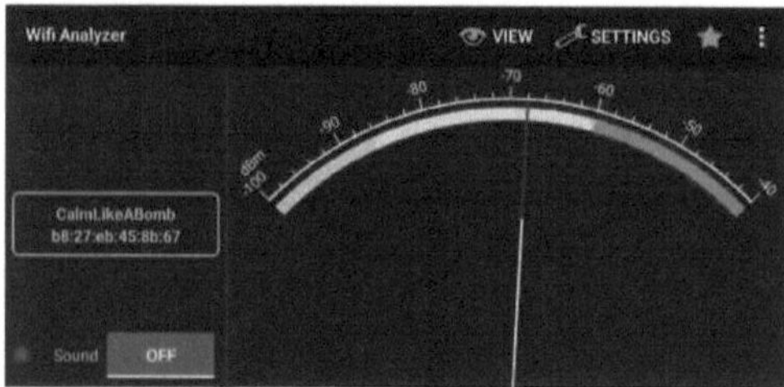

Figure 45 - Wi-Fi signal strength test
Source: The Author.

metres) than using Wi-Fi alone. However, the camera's images were unavailable at a distance of more than 25 metres.

3.7.5 Images taken from the camera

The images obtained from the camera in the supervisory system showed a slight delay due to the oscillations presented by the Raspberry Pi's Wi-Fi. Because the device is in constant motion, when it moves away the signal starts to slow down a lot, slowing down considerably and increasing packet loss. One way of improving this performance would be to install a router in the robot and another in the supervisory system, thus ensuring better signal quality and power.

In general, the images are of good quality and allow you to visualise details to the front and sides of the robot. It is possible to rotate the camera around 180° and check for obstacles to the sides and behind the robot, which is very helpful when you don't have a direct view of the equipment. Figure 46 shows one of the images obtained from the supervisory system during the tests carried out.

3.7. 6 Battery autonomy

For the battery autonomy test, the robot was set to move with the motors' PWM at maximum and timed to see how long it could keep the circuits functional under these conditions. The battery lasted approximately 23 minutes, knowing that the robot's speed is 1.33 Km/h, it would be able to cover a maximum distance of 522 metres before the battery ran out. This performance is compatible with the robot's function, taking into account that the maximum distance between the robot and the supervisory system cannot exceed 50 metres without losing control via the connection between the Xbee modules.

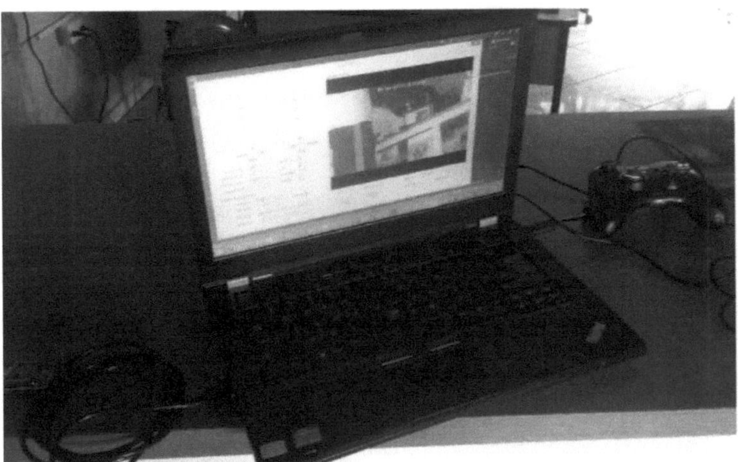
Figure 46 - Testing the Raspberry Pi camera and the supervisory system
Source: The Author.

3.7.7 Simulating an environment with explosive or flammable materials and toxic fumes

After making some adjustments to the positioning of the devices involved, calibrating the servo motors with maximum and minimum angles, as well as positioning the camera, a real simulation was carried out in which the robot accessed a location where there was a flammable device collected that was then triggered to emit heat and smoke, which allowed the robot's functionalities to be fully tested.

Figure 47 shows the image obtained from the supervisory system (A) identifying the object to be collected near 13kg Liquefied Petroleum Gas cylinders and the robot approaching the material (B).

Figure 48 shows the moment when the collected object begins to emit heat, increasing the temperature measured by the sensor and smoke begins to be released, also identified by the gas sensor (in A), the collected object being transported by the robot to a ventilated environment (in B).

The object is left in the ventilated environment, as shown in Figure 49, and begins to be completely consumed by the flames (in A), the operator performs an evasive manoeuvre to avoid damage to the robot that could be caused by the flames (in B).

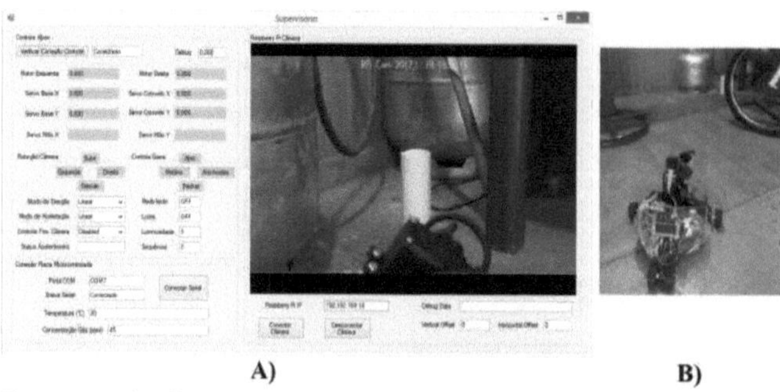

A)   B)

Figure 47 - Identification of flammable object to be collected
Source: The Author.

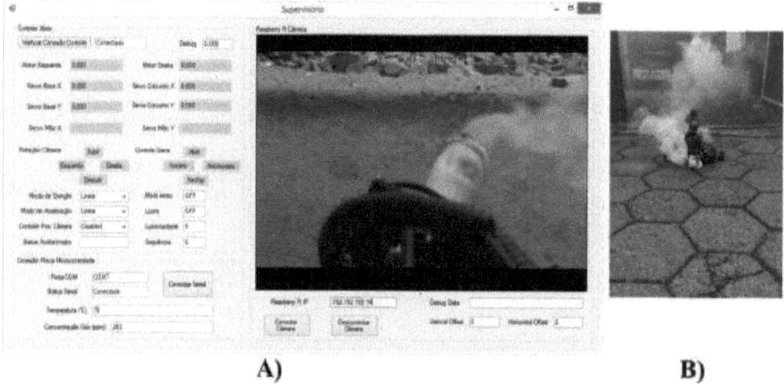

A)   B)

Figura 48   - Collecting the flammable object and moving it
Source: The Author.

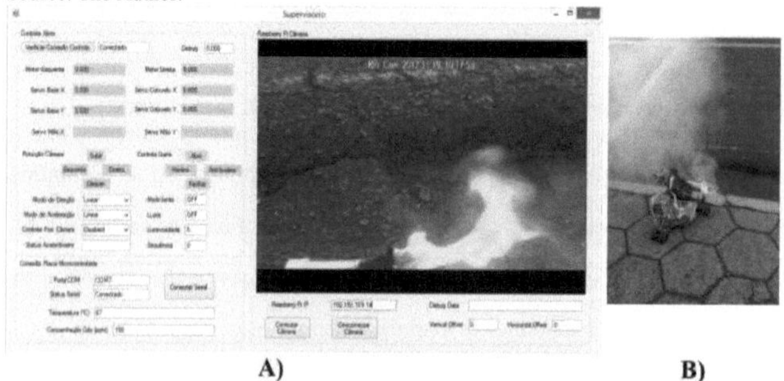

A)   B)

Figure 49 - Disposal of hazardous material in a ventilated environment
Source: The Author.

3.8 RESOURCES AND BUDGETS

3.8.1 *Hardware* features

As mentioned above, this project was developed using various *hardware* components. These include the Raspberry Pi board, a board developed with an ATMega2560 microcontroller, as well as a camera. In addition to these larger boards and devices, smaller, but no less important components were purchased, such as temperature and gas sensors, batteries, DC motors and servomotors, among others.

These features are included in the *hardware,* but as the components will not be purchased ready for use but will be built or adapted, it is important to mention them separately, namely the materials used to build the chassis, suspension, mechanical arm structure and acrylic sheets.

All the components were purchased by the academic, so the robot collector will be his property. Table 2 shows the main components that will be used, as well as the approximate value of each one (first half of 2017).

Table 2 - Budget for the components used

Costs			
Description	Quantity	Unit Value Average	Total
Acrylic (4mm sheet)	1	R$ 60,00	R$ 60,00
ATmega2560 microcontroller	1	R$ 40,00	R$ 40,00
Phenolite plates	3	R$ 60,00	R$ 180,00
Capacitors, resistors and other components	1	R$ 10,00	R$ 10,00
Camera for Raspberry Pi V2	1	R$ 150,00	R$ 150,00
Suspension trays	2	R$ 20,00	R$ 40,00
Batteries	16	R$ 5,00	R$ 80,00
Robotic arm	1	R$ 160,00	R$ 160,00
Chassis (sticks, glue, resin, fibre)	1	R$ 150,00	R$ 150,00
Springs and shock absorbers	4	R$ 35,00	R$ 140,00
Motors and gearboxes	2	R$ 80,00	R$ 160,00
Heat shrink PVC	1	R$ 40,00	R$ 40,00
Raspberri Pi 3 model B+	1	R$ 300,00	R$ 300,00
Wheels	4	R$ 20,00	R$ 80,00
Gas sensor	1	R$ 14,00	R$ 14,00
Temperature sensor	1	R$ 10,00	R$ 10,00
JX servo motors PDI-6221MG	1	R$ 150,00	R$ 150,00
Servo motors MG 995	8	R$ 30,00	R$ 240,00
Xbee S2 modules	2	R$ 120,00	R$ 240,00
		TOTAL	R$ 2.224,00

Source: The Author

CHAPTER 4

RESULTS

It can be seen that the robot can operate adequately according to its design, both in *software* and *hardware*. By adding another axis to the robotic arm and a transistorised H-bridge, it was possible to achieve greater precision when collecting materials, thus improving the robot's overall control. In the tests carried out in a commercial room with large openings, the robot's behaviour was as expected, managing to maintain communication with the supervisor at a maximum distance of 50 metres, allowing control via Xbee and detecting data from the environment such as temperature and the presence of toxic or flammable gases. Sending images proved to be stable up to a distance of 25 metres, with low frame loss and a delay of just a few milliseconds during transmissions. The autonomy of the robot was 23 minutes of operation, using the motors for locomotion at a maximum speed of 1.33 Km/h, it can travel a distance of up to 500 metres, without obstacles.

CHAPTER 5

FINAL CONSIDERATIONS

The microcontroller board developed proved to be functional, allowing the embedded *software* to be recorded and enabling all the ports selected for use, but its capabilities were limited to the development of the robot.

The embedded *software* developed for serial communication enabled the robot to be controlled satisfactorily, as well as reading sensors and using resources such as PWM.

The supervisory *software* developed for control proved to be suitable for the application, making it possible to manage both serial port and control events, as well as displaying images from the camera located on the Raspberry Pi.

It was necessary to modify the H-bridge, battery bank and add another axis to the robotic arm. These adjustments made it possible to control the robot's locomotion with variable speed and a greater range of motion of the robotic arm, making it easier to pick up objects.

The robot was tested in a commercial room with a flat floor and large openings and performed as expected, moving around to take readings from the gas and temperature sensors in the room. In 66 per cent of attempts, it managed to collect and transport the object.

The conclusion is that it is possible to develop a small, low-cost explorer robot. The battery autonomy time was 23 minutes, carrying a maximum load with the robotic arm of up to 150 grams. The values obtained by the sensors were transmitted at a maximum distance of 50 metres with a line of sight and the images from the camera were transmitted at a maximum distance of 25 metres using *Wi-Fi*.

REFERENCES

ADAMS, Herb. **Chassis Engineering: Chassis Design, Building and Tuning for High Performance Handling.** [S.L]: Federation of European Explosives and Manufactures, 1992. 144 p.

ATMEL. **ATmega2560.** [S.l.], 2017. Available at: <http://www.atmel.com/pt/br/devices/ ATMEGA2560.aspx>. Accessed on: 06.09.2017.

BENSON, Coleman. **History of Robots: Timeline.** [S.L], 2008. Available at: <www.robotshop.com/media/files/pdf/timeline.pdf>. Accessed on: 19.09.2017.

BRAUNL, Thomas. **Embedded Robotics.** [S.l.]: School of Electrical, Electronic and Computer Engineering, 2003. 458 p.

CAPUTO, Victor. **Meet the robot that will take care of security at the World Cup World.** 2014. Exame magazine. Available at: <http://exame.abril.com.br/tecnologia/ conhecera-o-robo-que-cuidara-da-segura-na-copa-do-mundo/>. Accessed on: 18.05.2017.

DEITEL, Harvey M.; DEITEL, Paul J. **C++ How to Programme.** [S.l.]: Pearson, 2006. 1208 p.

. **C How to Programme.** [S.l.]: Pearson, 2011. 846 p.

DIGI. **XBee and XBee-PRO RF Modules.** [S.l.], 2014. Available at: <https:
//www.sparkfun.com/datasheets/Wireless/Zigbee/XBee-Datasheet.pdf>. Accessed on: 09.10.2017.

FOUNDATION, Detroit Public Safety. **Bomb Squad**. 2015. Bomb Squad. Available at:
<http://www.detroitpublicsafetyfoundation.org/police-dpd/bomb-squad/>. Accessed on: 05.11.2017.

INOVATION, The Tech Museum of. **What is Robot?** 2005. The Tech Museum of Inovation.
Available at: <https://www.thetech.org/exhibits/online/robotics/activities/page02.html>. Accessed on:
13.08.2017.

INTEGRATED, Maxim. **DS18B20.** [S.L], 2017. Available at: <https://datasheets.
maximintegrated.encom/en/ds/DS 18B20.pdf>. Accessed on: 06.10.2017.

MACHADO, Claudinei. **20 Most Toxic Gases.** [S.L], 2014. Available at: <http:
//www.protecaorespiratoria.com/20-gases-mais-toxicos/>. Accessed on: 08.10.2017.

MANSOOR, Shamyl Bin et al. **Wirelles Bomb Disposal Robot**. 2001. Sir Syed University of Engineering
and Technology. Available at: <http://smansoor.seecs.nust.edu.pk/downloads/ WBDR_Report.pdf>.
Accessed on: 07.11.2017.

MARTINS, Gilberto de Andrade. **Manual for preparing monographs and dissertations.** [S.L]: Atlas,
2000. 298 p.

MAZIDI, Muhammad Ali. **Freescale ARM Cortex-M Embedded Programming: Using C Language**
[S.L]: Mazidi and Naimi, 2014. 682 p.

MICROSOFT. **Xbox 360 controller for Windows.** [S.L], 2017. Available at:
<https://www.microsoft.eom/accessories/pt-br/d/xbox-360-controller-for-windows>. Accessed on:
10.10.2017.

MIYADAIRA, Alberto Noboru. **PIC18 Microcontrollers. Learn and Programme in C.** [S.L]: Editora
Érica, 2013. 400 p.

MOLLOY, Derek. **Exploring BeagleBone: Tools and Techniques for Building with
Embedded Linux.** [S.l.]: Morgan and Claypool Publishers, 2011. 600 p.

MURPHY, Robin. **Introduction to AI Robotics.** [S.l.]: Massachusetts Institute of Technology, 2000. 466 p.

NASA. **Mars Exploration Rovers**. 2015. Nasa. Available at: <http://mars.nasa.gov/mer/ home/index.html>.
Accessed on: 18.08.2017.

PEREIRA, Fábio. **ARM technology 32-bit microcontrollers.** [S.l.]: Editora Érica, 2007. 314 p.

PI, Raspberry. **Raspberry Pi Model b+.** 2005. Raspberry Pi Foundation. Available at:
<https://www.raspberrypi.org/products/raspberry-pi-2-model-b/>. Accessed on: 11.08.2017.

RENO, Janet. **A Guide For Explosion And Bombing Scene Investigation.** [S.l.]: U.S.
Department Of Justice, 2000. 52 p.

RYDER, Chris. **A Special Kind of Courage.** [S.l.]: Methuen, 2005. 352 p.

SANDIN, Paul E. **Robots Mechanisms and Mechanical Devices.** [S.l.]: McGraw-Hill
Companies, 2003. 299 p.

SIEGWART, Roland; NOURBAKHSH, Illah. **Introduction to Autonomus Mobile Robots.**
[S.l.]: Massachusetts Institute of Technology, 2004. 321 p.

SKILLANO, Bruno; KHATIB, Oussama. **Springer Tracts in Advanced Robotics.** [S.l.]: Star, 2015. 510 p.

WASLAWICK, Raul Sidnei. **Analysis and Design of Information Systems Oriented to Objects.** [S.l.]: Elsevier, 2011. 329 p.

WISE, Edwin. **Robotics Demystified**. [S.l.]: McGraw-Hill Professional, 2005. 314 p.

APPENDIX A - COMPLETE CIRCUIT OF THE DEVELOPED COLLECTOR ROBOT

Presentation of the entire circuit used to develop the collector robot. Figure 50.

Figure 50 - Complete robot circuit

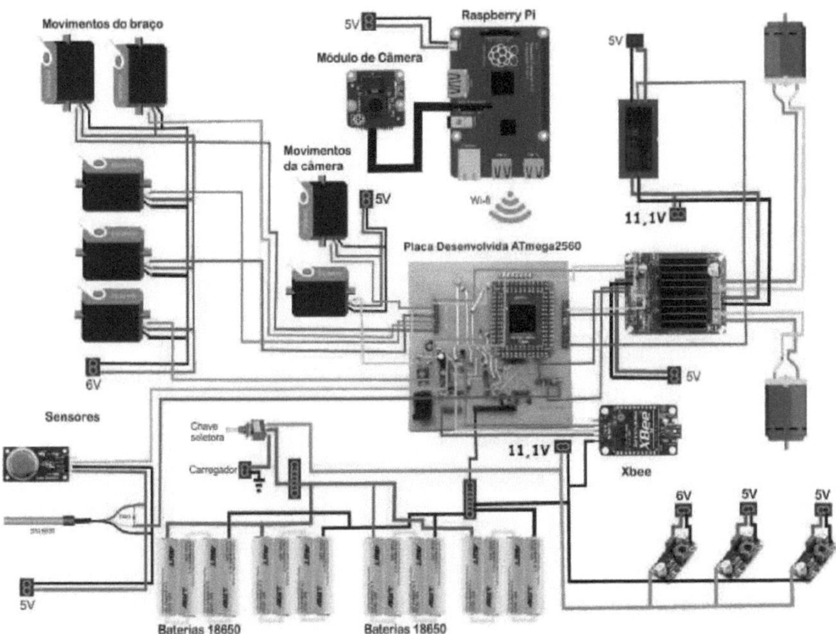

Source: The Author.

APPENDIX B - INITIALISING THE CONNECTION TO THE CAMERA VIA IP

```
private void cameraVvideoButtonClick(object sender, EventArgs e) {
Thread getVideoStream = new Thread(new ThreadStart(receiveVideoStream));
getVideoStream.IsBackground = true;
getVideoStream.Start(); }
private void stopVideoButtonClick(object sender, EventArgs e)
{ videorun = 0;
cameraImage.Image = Image.FromFile("defaultCamera.png"); }
int videorun = 1;
int upcommand = 0;
int downcommand = 2;
int leftcommand = 6;
int rightcommand = 4;
int horizOffset = 0;
int vertOoffset = 0;
private delegate void imageDelegate(Bitmap image);
private delegate void debugTextDelegate(string debugData);
private delegate void vertOffsetDelegate(string vert);
private delegate void horizOffsetDelegate(string horiz);
private void displayImage(Bitmap image)
{
cameraIimage.Image = image; }
private void displayDebug(string debugData)
{ debugTextbox.Text = debugData; }
private void displayVert(string vert)
{ vertOffsetValue.Text = vert; }
private void displayHoriz(string horiz)
{ horizOffsetValue.Text = horiz; }
```

```csharp
Thread updateIO = new Thread (new ThreadStart(UpdateSerial));
updateIO.IsBackground = true;
updateIO.Start();
private void UpdateSerial() {
while (true) {
if (portaSerial.IsOpen == true) {
if (received.Contains("A") && received.Contains("Z")) {
if (received.Contains("R")) {
rumblecount = 3;
}
serialIinputTextbox.Invoke(new inputDdelegate(displayIinput), received); }
}
}
else
{
serialOutputTextbox.Invoke(new outputDelegate(displayOoutput), "Sem conexão");
}
if (portaSerial.IsOpen == false)
{
Thread.Sleep(20);
}
}
```

APPENDIX D - SIGNATURE OF THE METHOD USED TO DETECT XBOX CONTROLLER EVENTS

```
Thread updatecontroller = new Thread(new ThreadStart(UpdateState));
updatecontroller.IsBackground = true;
updatecontroller.Start();
GamePadState state = GamePad.GetState(PlayerIndex.One);
sharedLeftTrig = state.Triggers.Left;
sharedLeftTrig = processRawThrottleData(sharedLeftTrig);
leftTrigValue.Invoke(new leftTrigDelegate(displayLeftTrig), displayLeftTrig);
sharedLeftStickX = state.ThumbSticks.Left.X;
sharedLeftStickX = processRrawSteerData(sharedLeftStickX);
leftStickXValue.Invoke(new leftStickXDelegate(displayLeftStickX), displayLeftStickX);
sharedLeftStickY = state.ThumbSticks.Left.Y;
leftStickYValue.Invoke(new leffStickYVdelegate(displayLeftStickY), sharedLeftStickY);
sharedRightTrig = state.Triggers.Right;
sharedRightTrig = processRawThrottleData(sharedRightTrig);
rightTrigValue.Invoke(new rightTrigDelegate(displayRightTrig), sharedRightTrig);
sharedRightStickX = state.ThumbSticks.Right.X;
rightStickXValue.Invoke(new rightStickXDelegate(displayRightStickX), sharedRightS-
tickX);
sharedRightStickY = state.ThumbSticks.Right.Y;
rightStickYValue.Invoke(new rightStickYDelegate(displayRightStickY), sharedRightS-
tickY);
```

Termo de Autorização para Publicação Eletrônica na Biblioteca Digital da

Universidade do Oeste de Santa Catarina – Unoesc

1. Identificação do material bibliográfico:

   (✗) TCC Graduação

2. Identificação do documento/autor

   Curso de Graduação: _ENGENHARIA DE COMPUTAÇÃO_

   Área de concentração (Tabela CNPQ) _1.03.04.00-2 SISTEMAS DE COMPUTAÇÃO_

   Título do TCC: _DESENVOLVIMENTO DE ROBÔ PARA EXPLORA-_
   ÇÃO DE AMBIENTES PERIGOSOS UTILIZADO SUPERVISÓRIO

   Autor _RAFAEL PABLO MASSOCATO_

   RG: _4755085_ CPF: _051 242 099-84_ Telefone _(49) 98844.1447_

   E-mail: _RAFAEL MASSOCATO @ GMAIL.COM_

   Orientador: _DANIEL CALIXTO FAGONDE MORAES_

   CPF: _365 253 750-53_ e-mail: _DANIEL.MORAES @ UNOESC. EDU. BR_

   Números de páginas: _74_

   Data de defesa: _18/12/17_ Data de entrega do arquivo: _21/12/17_

3. Informações de acesso ao documento

   Este trabalho é confidencial?[1] (✗) Sim ( ) Não

   Pode ser liberado para publicação na Biblioteca Digital ( ) Total (✗) Parcial
   Em caso de publicação parcial, assinale as permissões:
   ( ) Sumário
   (✗) Resumo
   ( ) Bibliografia
   ( ) Outras restrições:

   Na qualidade de titular dos direitos de autor da publicação supracitada, de acordo com a Lei nº 9610/98, autorizo a Universidade do Oeste de Santa Catarina – Unoesc, a disponibilizar gratuitamente, sem ressarcimento dos direitos autorais, conforme permissões assinadas acima, do documento, em meio eletrônico, na Rede Mundial de Computadores, no formato especificado[2], para fins de leitura, impressão e/ou download pela Internet, a título de divulgação da produção científica gerada pela Universidade, a partir desta data.

4. Está sujeito a registro de patente?
   (✗) Não
   ( ) Sim. Informar o nº do processo de encaminhamento ao Escritório de Interação e Transferência de Tecnologia.

   Local e data, _21/12/17_

___________________________ ___________________________
Assinatura do autor Assinatura do orientador

[1] Esta classificação poderá ser mantida por até um ano a partir da data de defesa. A extensão deste prazo suscita justificativa junto à Coordenação do Curso. Todo resumo estará disponível para reprodução, conforme Regulamento do Programa de Pós-graduação e Pesquisa da Unoesc.

[2] Texto (PDF); imagem(JPG OU GIF); som (WAV, MPEG, AIFF, SND); vídeo (MPEG,AVI,QT), outros (específico da área)

yes
I want morebooks!

Buy your books fast and straightforward online - at one of world's fastest growing online book stores! Environmentally sound due to Print-on-Demand technologies.

Buy your books online at
www.morebooks.shop

Kaufen Sie Ihre Bücher schnell und unkompliziert online – auf einer der am schnellsten wachsenden Buchhandelsplattformen weltweit! Dank Print-On-Demand umwelt- und ressourcenschonend produziert.

Bücher schneller online kaufen
www.morebooks.shop

info@omniscriptum.com
www.omniscriptum.com

MIX
Papier aus verantwortungsvollen Quellen
Paper from responsible sources
FSC® C105338
FSC
www.fsc.org